Beware of Car Hacking

A Systematic Analysis

Dipl.-Ing.(FH) Andy Gudera

tredition GmbH, Hamburg

If developers do not learn
to think destructively,
their peaceful products
can be used as weapons.

Special thanks to collegues and friends
for brainstorming and proofreading.

Cover: Andy Gudera

Verlag: tredition GmbH, Hamburg

ISBN Paperback: 978-3-7323-6368-1
ISBN e-Book: 978-3-7323-6369-8

I dedicate this book
to my daughter
so that she can grow up
in a safe and joyful world
of automobiles.

Contents

In the last few months a number of eye opening car hacks[1] became known in daily press and trade press. These attacks on a dearly beloved object of our culture and freedom represent the changes of technology and human behaviour in an impressive way.

This book should be read as a first rough analysis of hacking motivations, the people behind, potential impacts and technical backgrounds, based on many years of experience by working with several manufacturers and suppliers in the fields of vehicle electronics, especially questions of safety relevant questions of chassis electronics, e-mobility, vehicle network, diagnostics, driver assistance systems and functional safety.

Many options and methods described hereafter are far more theoretical in nature right now, but they have a high potential to be used for attacks on cars.

Although this discourse may seem to be an instructions guide to hack vehicles, its aim is to identify weaknesses by systematic analysis to close potential loopholes effectively. Likewise, it is a tool for the development of safe and secure vehicles.

Fig. 1.: DIY remote control [17]

[1] demonstration of remote control steering and brakes [28], simulated burglary via online service of a car [7]

Part I.

Introduction

Since 1886 cars have been moving on our roads. The initial scepticism about the safety of the rattling and stinking monsters has rapidly turned into a widespread acceptance in all walks of life.

The comfort, convenience and personal freedom of individual mobility outweighed concerns about the risks. Over time, the technological progress has led to reduced accident rates and their consequences.

In this state of fundamental confidence in technology suddenly warning calls concerning new risks arose. The automotive industry with its mechanical roots was shocked and taken by surprise.

Finally, the software attacks exposed the immature state of defensive software development in automotive products.

What happened? What circumstances have made these attacks possible?

Fig. 2.: Elwood Haynes driving his first automobile in 1886
photo by M. C. Madden

This part aims to give a first impression why the topic of hacked cars has become such an issue right now.

Therefore it is necessary to analyse technical trends as well as the presumed motivation of hackers.

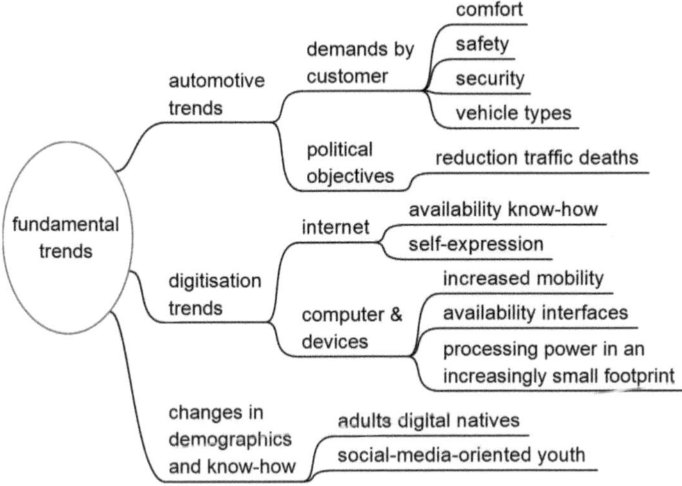

Fig. 3.: fundamental trends for hacking

As the mindset of the manipulators are widely spread, the claim of completeness can of course not be given.

The bandwidth of possible hackers is also enormous. Their motivations and understanding are very different in nature. A brief overview is given in Figure 4.

Fig. 4.: hacker types

Finally, this part is about the **WHY** right now and **WHO** are the involved persons.

1 Automotive Trends

In recent years automotive electronics have been developing at a rapid speed. This can be explained simplified by reduction on two major topics: general trends in the automotive industry and the increased availability of powerful automotive embedded systems on the market.

Fig. 5.: reflections on vehicle (r)evolution
photo by author

1.1 Overview Automotive Trends

Changes in the automotive market in terms of customer requirements as well as political objectives are evident. Several reasons for this (r)evolution of the automobile world can be identified; including, but not restricted to,

- increasing demand for comfort by the customer
- increasing demand for safety by the customer
- increasing demand for SUV[1] by customer
- increasing demand for personalisation of the car by the customer
- demographical changes of customer behaviour - more and more driver's are digital natives
- strong ambition in politics to reduce fuel consumption and CO_2 emission up to zero emission
- politics want to reduce traffic deaths
- pricing pressure[2] for OEM and supplier
- generation of unique selling propositions[3] (USP) by OEM
- demand of cost reduction in workshops by increased diagnostics
- increasing diagnostics due to increasing complexity of vehicle systems

Further reasons can be added to this list easily.

To fulfil all of these demands, the number of electronic controller units (ECU) has increased up to over 60 in premium vehicles.

1.2 Special Trends - Curse or Blessing

To fulfill the customers demands, postulated or supposed, a number of applications are in development at most OEM.

1.2.1 Autonomous Driving

The most prominent example is the autonomous driving. OEMs and suppliers are competing for interesting solutions that are both commercially feasible as safe from functional point of view.

[1] sports utility vehicle
[2] often driven by shareholder value
[3] first to market, technical leadership

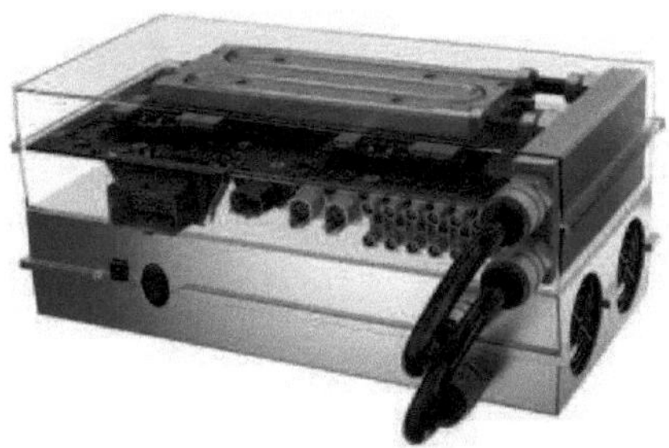

Fig. 6.: Nvidia Drive PX 2 supercomputer for autonomous driving [16]
photo by Nvidia

Not only established automotive suppliers take part at the race for technology leadership in new trends, such as autonomous driving, but also electronics manufacturers from infotainment and multimedia branch.

A good example is the Nvidia Corporation, well known for graphic controllers and gaming consoles. Since a few years they engage increasingly in the automotive sector, including in the self-envident field of graphical instrument cluster.

The latest step of Nvidia was to unveil a highest performance supercomputer for autonomous driving using artificial intelligence [16].

Fig. 7.: Nvidia Drive PX 2 - application using artificial intelligence [16]
photo by Nvidia

As already expected at the suppliers with new players, there are also at the OEMs new participants, such as small manufacturers[4] and other automotive underdogs[5]. Interestingly enough the underdogs seems to have a clear lead to the big players.

However, that this assessment be seen critically, showing abnormalities at Tesla S autopilot [39]. It is fascinating, that Tesla customers as early adopters despite clearly recognizable hazards tolerate these fails.

The public outcry for such failures in the big brands would hurt these companies significantly more. Series production of autonomous vehicles on a large scale can only be achieved at least in compliance with various standards[6] including development rules for OEM and supplier and testing activities from idea to ready product. Pretty rushed hardware and software would be downright negligent here.

---

[4] e.g. Tesla Motors
[5] e.g. Google, Apple
[6] e.g. ISO 26262

Artificial Intelligence

That the classical programming pushes to their limits while autonomous driving, should be clear with a short view on the increasing situation catalogs which have to be implemented by program code.

Such as the human driver as to learn to assess situations in the traffic, machine learning[7] for vehicles seems to be a chance to obtain the complexity of driving under control.

A trend this already shows the Nvidia Drive PX2 supercomputer [16]. But even Toyota seems to want to implement autonomous driving via artificial intelligence [15] and is investing heavily in this area [40].

Even if artificial intelligence certainly constitutes a technical solution for autonomous driving, the programmatic blur of self-learning systems will pose several questions for standardisation[8], testing and the legislature.

Car-to-Car and Car-to-X Communication

Autonomous driving means for the vehicles computer to be able to process as much data on the environment as possible.

An idea to use not only the vehicle's sensors and information sources, the Car-to-X communication. This technology allow the car, to use data which have been collected from other vehicles, but also data of stationary transport facilities[9]. The preferred communication medium is wireless lan (Wi-Fi).

Whether it will however be in this technology to curse or a blessing, the future will tell. That a deliberate use and caution will be necessary should be shown in 10.3.4 Wi-Fi and 11.1.4 Vehicle Network Architecture.

---

[7] artificial intelligence (AI)

[8] ISO 26262

[9] e.g. traffic lights, signages

1.2.2 Design Optimisation

A waiver of exterior mirrors at first appears rather than gimmick than real need.

Fig. 8.: mirrorless BMW i8 [9]
photo by BMW

But a closer look on the aerodynamic mirror replacing cameras can explain their usage.

In times of reduction of CO_2 emissions and limited range of electrical vehicles an optimisation of aerodynamics make a lot of sense.

Fig. 9.: display BMW i8 mirrorless [9]
photo by BMW

That this technology does have several potential risks[10], but should be kept in mind.

Displays are very popular as design elements suggesting diversity of information. Similarities to the popular sci-fi film sets and scenes are there not by chance.

[10]potential manipulations of camera module, video stream and display content

Fig. 10.: oversized screens at BMW i8 [8]

As futuristic large screens may look like, they are not without risks in different ways[11].

1.3 Electronics vs. Mechanics

Many systems, previously implemented by using mechanical principles, have been enhanced and transformed by electronics, such as steering systems[12], damping systems[13] or parking brakes[14].

The degree of computerisation differs from the traditional understanding of mechanical engineering of many automotive OEM. A fundamental change of the relationship between electronics and mechanics occured.

[11] see subsection 4.2.1 - Eyesight Effects on page 45

[12] hydraulic power steering replaced by electric power steering [19]

[13] conventional damping systems supplemented by electronical air suspension [2], dampers with electrically influenced viscosity [5]

[14] brake cable replaced by motor driven brake callipers [18]

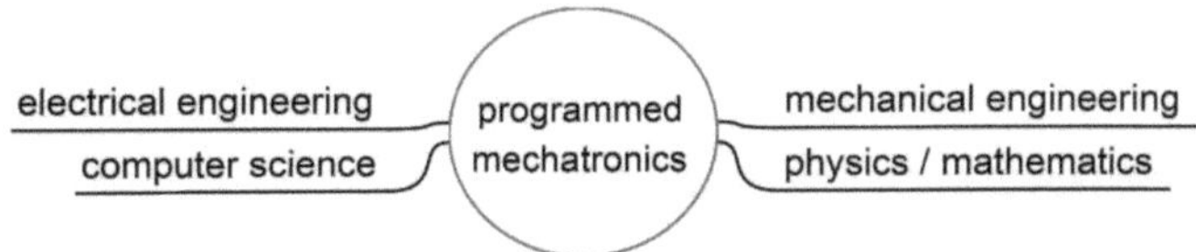

Fig. 11.: cross-divisional function mechatronics

The use of electronics containing hardware and software in mechanical systems requires new ways of thinking, methods and processes. It's not about stirring up competition and rivalry between mechanics and electronics, but finding a new way of ***programmed mechatronics*** in the automotive world. The car manufacturers urgently need to establish interdisciplinary teams of experts to address the hacking threats.

The fruit of all these considerations is ultimately still an automobile. This is true for machine builders as well as for electronic engineers and programmers. That should never be forgotten.

2 Digitisation Trends

In addition to the automotive trends, a furious progress in digitisation results in a fundamental change of the situation in the automotive market, which have to be analysed separately.

The impact of the digitsation on the vehicle is mainly based on two parts of computer technology - evolution and diversification in hardware devices and the networking of them, because hackers will target cars[1] as the next attack vector.

2.1 Computers and Devices

First, to something tangible. The market of hardware and software for personal computers and related devices has undergone significant changes in recent years, too.

A comprehensive analysis[2] of these changes in the market without blinkers[3] has proved to be absolutely necessary and fascinating.

2.1.1 Computer (PC)

Performance of current computers are demonstrating impressively the fact that Moore's Law[4] is still valid. That clunky desktop[5] computer models will be more and more replaced by stylish mobile computer like laptops[6], ultrabooks[7] or netbooks[8], does not change the significance of the personal computer. The PC is still the all-purpose machine for electronic data processing.

---

[1] car as part of Internet of Things (IoT)

[2] technology scouting

[3] also considering improper use of devices

[4] The number of transistors in integrated circuits doubles every two years. [44]

[5] mainly used for server or gaming computer

[6] usually large screens, high performance

[7] stylish, weight- and energy-optimised low power CPUs, slim, high mobility

[8] price optimised, small screens, low power

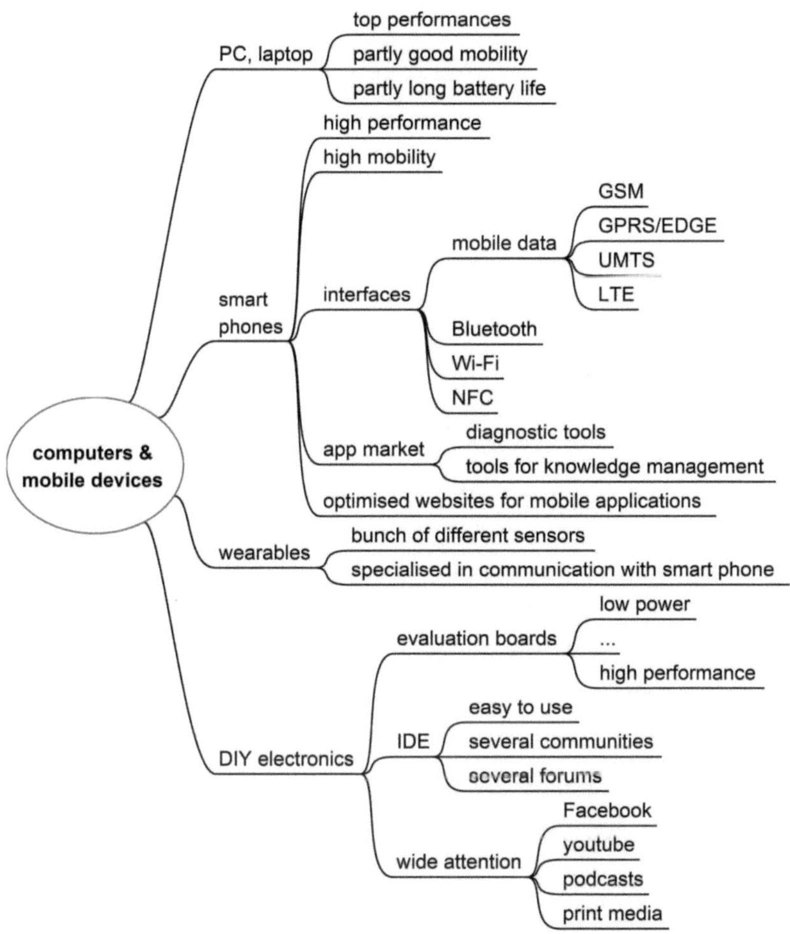

Fig. 12.: aspects of computers and devices

The performance of an solitary computer is no longer that important nowadays. By now broadband network is highly available and allows to use computing power in the cloud. Besides, the prices for personal computers have fallen significantly.

Similarly, the software market clearly shows changing payment models[9,10] that make previously extremely expensive software interesting and available for the ordinary consumer.

Widespread availability of hardware and software products also led to an increased knowledge about it.

2.1.2 Smart Phones, Tablet Computers

Smart phones, phablets[11] and tablet computer[12] - all based on SoC[13] - are taking over more and more tasks of the desktop or laptop computer.

That's not a contradiction to subsection 2.1.1. Actually, this involves additional user groups that previously never would come into contact with computers such as little children and the elderly. The intuitive devices have not only changed the age limits of users. They also reduce the required knowledge for usage of those devices dramatically.

Entertainment[14], navigation and recreation applications are highly available and increase the demand of these features in cars, and not only for technically minded customers.

This trend is also interesting for the automotive industry.

A couple of new releases of leading automotive manufacturers are already showing applications for smart phones and smart watches in combination with vehicles[15], the latest technology at the moment.

[9] rent instead of buy

[10] e.g. Adobe Creative Cloud, Microsoft Office 365

[11] smart phones with larger screens

[12] usually without telephony functionality

[13] System on Chip, multi functional microcontroller for smart mobile devices

[14] music, video, games

[15] e.g. advanced parking assist, power management for electric vehicles

2.1.3 Wearables

Wearables are computer devices or extensions that can be worn directly on the body of the user. The range starts from fitness or activity tracker to smart watches[16] up to head mounted displays like Google Glass or virtual reality (VR) glasses[17].

Fitness Tracker, Activity Tracker

The usage of wearable health sensors for driver monitoring[18] could be quite interesting in the vehicle. This is certainly only a matter of time[19].

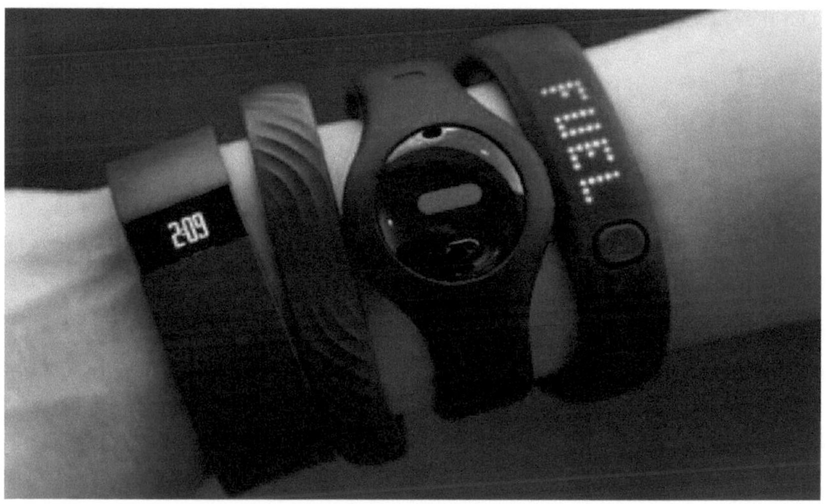

Fig. 13.: several fitness tracker [26]
photo by Richard Drew

[16] see Figure 2.1.3 - Smart Watches on page 21

[17] e.g. Oculus Rift, Samsung Gear VR, HTC Vive

[18] driver drowsiness detection, cardiovascular monitoring

[19] "Audi Fit Driver" at CES 2016

Wearables like fitness or activity tracker are highly integrated sensor clusters based on performance and price-optimised microcontrollers[20] that are proprietary developed and programmed.

Typical implemented sensors are

- puls rate
- oxygen saturation
- 3D acceleration sensors
- temperature sensors[21]
- respiration rate
- GPS
- UV sensors[22]
- ambient light sensors[23]
- microphone[24]
- galvanic skin response sensor[25]

Not only the equipment of sensors but also the designs are very different. There are bracelets, pendants, with or without display, vibration alarm etc.

Typically the communication is based on Bluetooth. But not only Bluetooth communication represents a potential vulnerability[26]. The development and maintenance[27] of wearables offer potential weaknesses.

Only a relatively small team developes the software, that has to be realised for reasons of memory space and performance optimisation mostly without the common operating systems. So bugs and security lacks remain for a longer time in the unit and open the door to hackers.

[20] ASICS
[21] skin, environment
[22] sun burn protection
[23] day/night detection
[24] phone speaker
[25] sweat detection
[26] see subsection 10.3.3 - Bluetooth on page 118
[27] availability of software updates and bugfixes

Smart Watches

A quite new functional wireless[28] extension of smart phones on the gadget marked is represented by smart watches that also work partly autonomously.

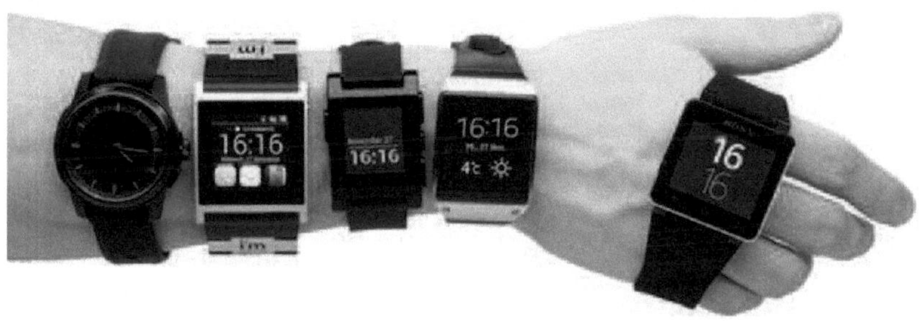

Fig. 14.: several smart watches [36]

Small touch screens with acoustic and vibration alarms make trendy human-machine interfaces possible, which are quite interesting for applications in combination with cars. But there are interesting fields of manipulations[29,30], too!

In contrast to the powerful smartphones these devices usually have reduced performance to make small units and long battery lifetime by low power consumption possible.

[28] using Bluetooth or Wi-Fi

[29] see subsection 4.2.3 - Sense of Touch on page 52

[30] see subsection 4.2.4 - Distraction on page 53

Fig. 15.: Oxy Smartwatch, open source [34]

That security features are sacrificed to performance optimisations, is not inconceivable.

Quite interesting are smart watches based on the open source idea such as Oxy Smartwatch[31] [34], where the complete platform[32] [24] is available for modifications.

After all, smart watches rely in most cases on classical operating systems such as Apple iOS, Android, TIZEN & Co. leading to a broad community of developers. Therefore, it can be assumed that the developer community[33] can quickly identify and fix security holes.

Head Mounted Displays

Virtual reality is no vision for the future or military applications anymore. Head up displays (HUD) are on the option list of many vehicle manufacturers. So there is no big step from head up display including augmented reality [27] to head mounted display [10].

[31] see Figure 15 - open source smart watch on page 22

[32] Ingenic Wearable Open Platform

[33] The wisdom of crowds. [54]

Fig. 16.: augmented reality by Jaguar head up display [27]

Fig. 17.: concept for augmented-reality driving glasses [10] by BMW

Fig. 18.: head mounted display (BMW goggles [21])

Presumably, the legislature will only permit the transparent head-mounted displays a la Google Glass, but the possibilities are quite attractive. "Transparent walls", marking on distant accident sites or road damages, maybe even games can be implemented with the help of those wearables. Correspending risks are comprehensible[34].

[34] see subsection 4.2.1 - Eyesight Effects on page 45

Fig. 19.: helmet mounted HUD [11]
photo by BMW

Fig. 20.: helmet mounted HUD view [11]
photo by BMW

A clever implementation of VR displays in eyewear as a fashion accessory can minimize the recognisability of this technology for executive authorities dramatically.

2.1.4 DIY Electronics

A very interesting trend can be observed in the Do-It-Yourself (DIY) market.

Evaluation boards[35] with relatively low performance used to be exorbitantly expensive. Now there are inexpensive single-board computers in all performance classes and several equipment[36] available for little money, including easy to use development environment and supporting forums and newsgroups.

[35] development board containing a microcontroller and minimum of supporting environmental components; technical basis for low-level software development
[36] e.g. BeagleBone board [6] including CAN bus [47]

Fig. 21.: Raspberry Pi with additional CAN bus shield (CarBerry) [13]

The range starts from from 8-bit Arduino [4] controller boards with application specific shields[37] and lightweight preconfigured IDE[38] and ends with high performance System on Chip (SoC) ARM[39] processors[40] comparable with actual smart phones and several available embedded operating systems[41].

Not only the low price, a good handling and a lot of available extensions (shields) but also the broad support by forums and communities expand the application options of these products significantly.

In addition, through social media[42], print media, podcasts etc. these devices and their (hacking) applications reach to a wide audience.

2.2 Internet

Some of the current trends have led to a significantly change of use of the Internet. The way people use it and the content they can access emerged from closed groups of Internet gurus to a common behaviour of persons of all classes.

[37] add on board with special interfaces, displays etc.

[38] integrated development environment; platform for programming and debugging of software

[39] Acorn RISC Machines, controllers with RISC (Reduced Instruction Set Computer) architecture

[40] Raspberry Pi [35], [47], BeagleBone, UDOO DUAL & QUAD [38], ODROID boards [33]

[41] embedded Linux, Android [3], GENIVI [23], TIZEN [37] etc.

[42] e.g. Facebook, youtube

2.2.1 Mobile Internet

Unleashed from the home computer and dark basement hobby rooms the Internet starts to be mainly used on mobile devices such as smart phones or tablet computers.

This increasing mobility always allows the acquisition of more knowledge[43], leisure[44] and communication[45] - any time, anywhere.

2.2.2 Availability of knowledge

The Internet has helped the "The Wisdom of Crowds" [54] to a completely new quality. Hackers can share their knowledge instantaneously on any device, i.e. boasting of hacking or hacking skills.

Systematically online encyclopaedias[46] [43] have been established, such as digital and digitised collections of books and gained application-related information in forums.

In combination with the already mentioned mobility of the Internet, this knowledge can be retrieved spontaneously.

2.2.3 Social media and visibility

Social networks such as Facebook create a high pressure of ostentatious self-realisation and public thrills as well as the sharing of experiences.

---

[43] spontaneous query of knowledge (e.g. Wikipedia, Google search)

[44] multimedia (audio and video streaming), social media

[45] phone calls, instant messenging

[46] e.g. Wikipedia

Fig. 22.: small selection of social media and platforms

Further on, the social media represent an experience sharing platform leaving a paper-trail of clues for investigators and car manufacturers for learning how car hackers behave.

3 Motivation of Hacking

As a result of the digitisation trends a certain number of motives for hacking emerged.

Hacks generally spoken are merely manipulations on hardware and software systems that are not intended by the manufacturer of a product regardless of their intention.

Two main streams of hacking should be discussed separately: non-destructive and destructive manipulation of automotive electronics.

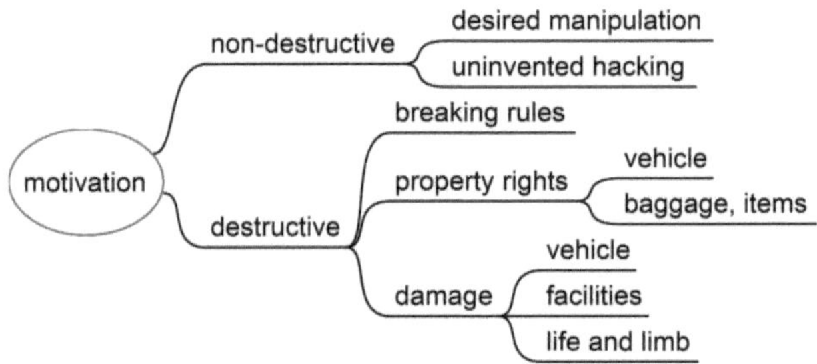

Fig. 23.: motivation of hacking

3.1 Non-Destructive Hacking

Not all manipulations have to be expected to be harmful, although the hack exceeds system values of the vehicle, which are planned, defined and tested by the manufacturer.

The change of system states on planned boundaries occasionally can even have a positive impact on the vehicle[1,2].

[1] optimised fuel consumption

[2] see Figure 24 - eco-tuning by Carlsson on page 31

Nevertheless discovered manipulations go hand in hand with a loss of warranty.

3.1.1 Desired Manipulations

As already mentioned the customers wish of individualised and personalised vehicles asks for non-destructive[3] hacking.

A typical case of non-destructive hacking or modding[4] is the mostly commercial car tuning scene.

Car tuning companies are modifying engine power, behaviour and speed of automatic drives etc. Therefore they act (mostly) with a lot of technical expertise to implement the tuning measures.

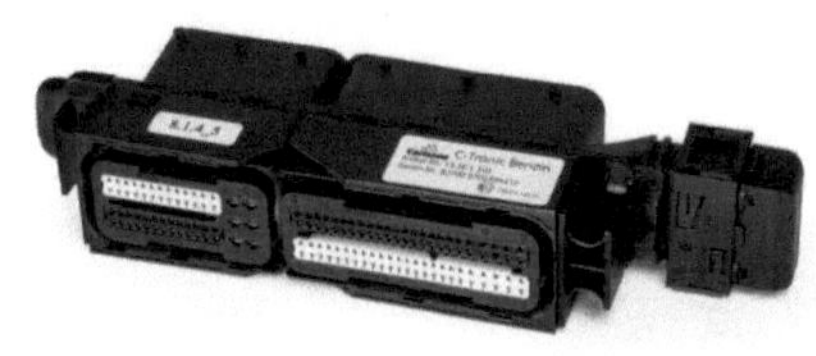

Fig. 24.: eco-tuning by Carlsson C-TRONIC® MODULE [14]

In other cases of non-destructive hacking, car owners try to adjust functionalities[5] intended by the manufacturer exceeding the given limits.

Desired hacking is often realised by usage of standard tools[6] from OEM or workshop without external penetration.

Typically an impact on driving stability and safety is not given.

---

[3]user friendly modifications of behaviour and appearance

[4]mainly changes in appearance

[5]changing colors of ambiant lightning, intensity of screen backlights, replacement of default screens of infotainment systems

[6]diagnostic tester

3.1.2 Unwanted Hacking

Not in every case changes of car behaviour are desired by the customer, especially if the changes are the result of external attacks.

Uninvited hackers commonly use unconventional interfaces and not intended ones[7].

Fig. 25.: DIY gateway - Samy Kamkar's "RollJam" device
photo by Samy Kamkar

Depending on the 'polluter pays[8]' here is more of a playful aspect to suspect, but systematic attacks[9] are increasingly likely.

A systematic attack by someone[10] to discredit carmakers can also not be ruled out.

3.1.3 Violation of Privacy and Security Policy

Are the car manufacturers even aware that they themselves can also be counted among the hackers?

[7] Wi-Fi, Bluetooth, GSM, USB, SD cards
[8] script kiddies, bored Internet gurus
[9] mafia structures, secret services
[10] e.g. frustrated or corrupt programmers at the ECU suppliers

The systematic tapping of data that do not serve the intended use but the analysis of customer behavior, is also very close to the limits of legality as hacking by outsiders.

Revelations of unexplained data usage at BMW by Fédération Internationale de l'Automobile (FIA) [22] shows, that this behaviour is no fiction anymore.

However, it should not be an isolated case. Evidence of misuse of data relating to driving behaviour with other manufacturers is surely only a matter of time.

To impute intent, would certainly be premature. Rather, it can be assumed that it is actually here overshot targets. An understanding of the usage and misuse of data seems to be not arrived yet at all areas of the automotive industry. But ignorance is no excuse, however.

3.2 Destructive Hacking

Quite the contrary to the purposely "modding" of the car without wrecking[11] is the destructive hacking.

The main difference to non-destructive hacking is the violation of legal and moral rules partly in conjuction with wrecking the car or its components.

3.2.1 Breaking Rules

A typical case of breaking rules is the deactivation of safety mechanism.

While the switching off of belt warner is still rather harmless, the deactivation of TV screen locking already constitutes a gross violation of the law.

Also the lowering of distance limits of adaptive cruise systems below the limits fixed by law is breaking rules. A particularly serious violation

[11] no targeted damages for car or components

of applicable law is the manipulating mileage counter[12] readings[13] of vehicles.

Usually an immediate danger to life and limb is not given.

3.2.2 Violations of Property Rights

Much more impact on car owners have hacks on the car security such as entrance systems or immobilizer.

Granted, in some cases[14] thefts have an easy time. These systems are the best example for the conflict between convenience and security.

Not in every case the complete car is the target of thieves. Sometimes the trunk content, visible equipment[15] electronic[16] or interior[17] parts are much more interesting than a complete car with a lot of trackables[18].

Violations of property rights are directed by the expected profit as well as cost efficiency. Therefore a burglary with the help of electronics only makes sense if revenues exceed the expenses.

These individual offences are ruled out as usual. The technical effort is crying out for multiple use and thus tied moderate theft.

3.2.3 Damage to Property

A very different motivation is the damage of property. Behind this subject various objectives are hidden.

Damage of the Car

A first possible way of damaging a car would be the deactivation. At first glance, it sounds funny. It becomes critical if the action without

[12]instrument cluster

[13]falsification of documents

[14]remote keys, remote keyless entry systems

[15]tools, measurement equipment

[16]navigation systems, airbags, clima controller

[17]seats, belts, dashboards

[18]serial notes, immobilizer

any special intentions is not a silly youthful prank, but clearly targeted attack by mafia structures or secret services. Organisations such as these fear no technical and financial effort.

In these cases, it is no longer about mischief.

Also conceivable is blackmailing against automobile manufacturers. Activities like these don't need a big team of criminal individuals. It is already enough if there is a single, frustrated employee[19] who is willing to annoy a company.

Damage of Road Traffic Facilities or Buildings

Far from children's games, cars are being used to attack traffic facilities or buildings.

The use of cars for such attacks suggests that no consideration is given to life and limb of the occupants.

3.2.4 Damage to Life and Limb

It is finally not a long way from destruction to murder. A special quality of such attacks are that no or only few traces are left. In this way almost no suspicion is aroused!

Perhaps a part of the behavior of Stephen King's Christine[20] would so at least explain. Imagination in 1983, reality in 2016?

3.3 Summary - Hacking Motivation

The range from minor desired modifications, upward to annoy people and companies to premeditated murder using car electronics is quite large and has to come to view.

Knowing that there are people with motivations like this a special consideration in the design of automotive electronics must be given.

[19] with sufficient knowledge of technology and processes

[20] a red and white 1958 Plymouth Fury which kills quasi autonomously several people

The usefulness of functions and resulting risks and damages must be also weighed by non-technical organisation units such as marketing, design, legal and financial risk assessors.

Part II.

Symptoms of Hacking

After the clarification of **WHY** and **WHO** in this section the **WHAT** with its scope on the driver will be the main topic.

It is intended to shed light on the symptoms, which are caused by hacking. A special overview on their impact on the people in and around the vehicle shall be given.

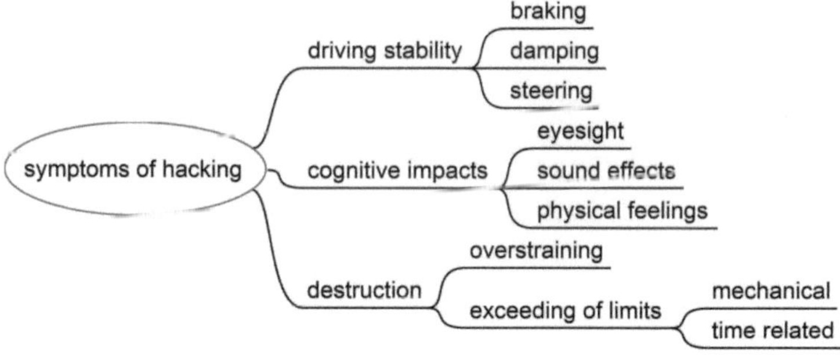

Fig. 26.: symptoms of hacking

4 Kinds of Manipulation

First a few thoughts on the kinds of manipulation in detail; which options do hackers have in principle.

It is not clear from the beginning, whether or not the manipulation is destructive. In most cases it depends on the hacking effects severity and can vary from mild to fatal. In assessing the impact of the hack the driver's skills and ability to influence are not insignificant.

4.1 Driving Stability

The impact of hacking on driving stability is quite easy to understand.

One of the main reasons driving instabilities would have a very hard impact is the presence of stability improving systems in current vehicles.

Fig. 27.: driving stability
photo by author

The availability of stabilising systems like ABS[1] or ESP[2] is accustomed to most types of driver's[3].

4.1.1 Braking

Braking systems have a very direct impact on the driving behaviour of the vehicle. That is why a special attention to those systems is urgently needed.

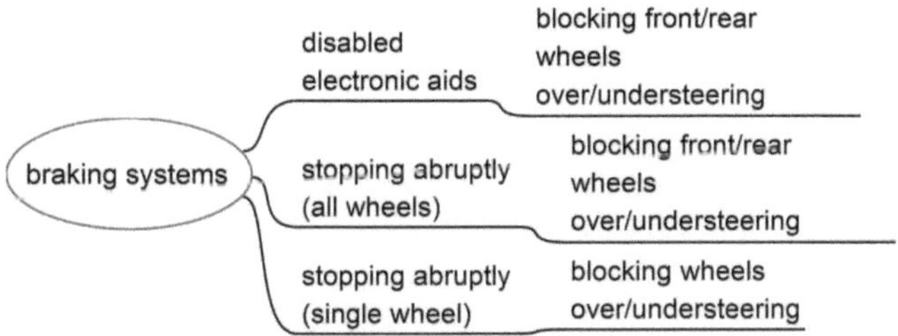

Fig. 28.: impact braking systems

Since braking systems affect each wheel of a vehicle individually, they can also achieve steering effects in addition to linear decelerations, such as oversteering and understeering.

Oversteering, Understeering

Many driver's find manoeuvres of oversteering difficult to handle. The rear axle wheels loose their traction and the rear part of the car slides. Of course, an inexperienced driver will find this manoeuvre difficult to control. Even for the experienced driver oversteering is not an easy task.

In fact most car builders try to limit this vehicle behaviour. Therefore the cars are adjusted for a manoeuvre easier to control, the understeering.

---

[1] antilock braking system

[2] electronic stabilisation program

[3] from beginners to normal and experienced driver's, up to driver's with racing background

The tuning of the vehicle into certain behaviour by purely mechanical measures is complicated and expensive. The use of electronic aids reduces these costs significantly.

In some cases, manufacturers almost exclusively rely on electronic aids. If these systems fail, the resulting driving behaviour is unpredictable, also in rather uncritical situations[4].

Objective Braking Systems

Braking systems have a very immediate effect on the vehicle and the ability to drive it.

A single manipulated valve of a brake system may have catastrophic consequences. What a perfidious efficiency!

Therefore, attacks on electronic brake systems are very urgent and must be foreseen by car developers and prevented accordingly.

4.1.2 Damping

Mainly for luxury and sports vehicles electronic damping systems have been developed.

The impact defective dampers have on the driving behaviour, appears usually only on critical ground and faster-paced driving manoeuvres.

The loss of traction leads to already described under- or oversteering.

Objective Damping Controls

Attacks via damping controls have a more random effect and thus complicate predictability[5], repeatability and cause determination[6].

[4]lower speed, dry roads
[5]also for the attacker
[6]accident analysis

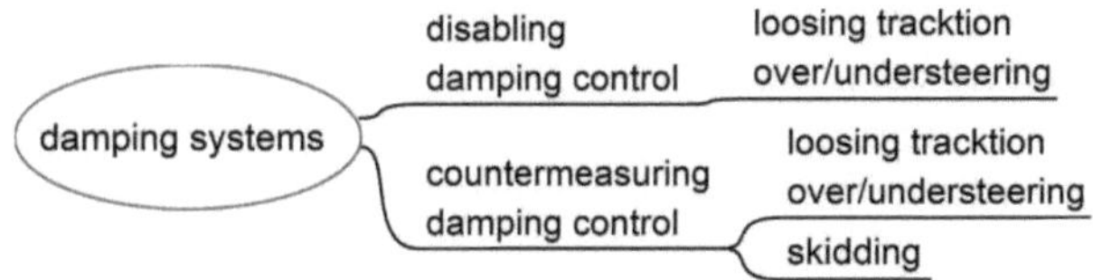

Fig. 29.: impact of damping control

Manipulation of damper systems - whether desired (Tuning) or not - can affect the performance of the vehicle dramatically and reduce the controllability.

4.1.3 Steering

Driven by cost pressures and a generation of USP[7], power steering systems were electrified[8] also for high vehicle classes with heavy engines.

This type of steering assistance allows new features such as the autonomous parking. That means the car drives and steers itself into a parking position without manual interventions by the driver.

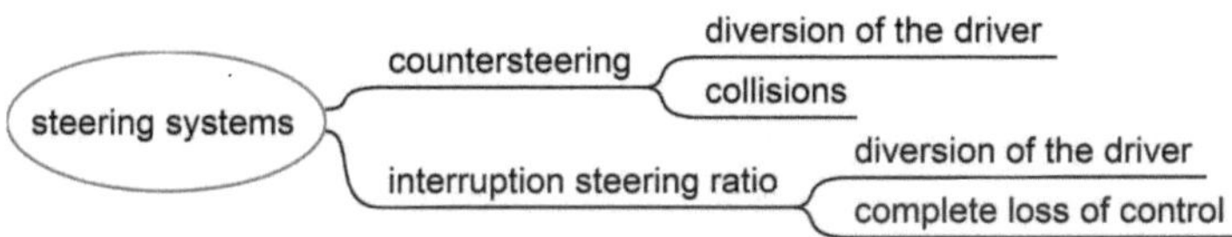

Fig. 30.: impact of steering aids

Further functionalities based on electrical power steering systems are driving steering recommendations[9], which also effect the steering wheel[10].

[7] unique selling proposition, unique selling point

[8] previously hydraulic solutions

[9] in case of understeering or oversteering the system supports the driver to countersteer

[10] see subsection 4.2.3 - Sense of Touch on page 52

Parking manoeuvres and driving at top speeds can be simplified by usage of steering systems with variable gear ratio.

Objective Steering Systems

Steering systems open up a wide field of possibilities for attacks. It ranges from minor vibrations[11] of the steering wheel to powerful steering intervention[12].

A mechanical separation[13] between steering wheel and steering system is also possible. In this case no fall back strategies will have an effect.

4.1.4 Powertrain

Not only the chassis electronics have a strong impact on driving stability, but also the powertrain containing driving engine[14], (automated) gear and 4wheel drive systems[15].

Especially high performance engines[16] are often limited in their effect[17] by driving aids for increased driving stability by the OEM. In combination with manipulated driving aids attacks on the powertrain would have a big impact.

Advanced 4wheel systems can distribute torque not only between front and rear axle, but between right and left side depending on traction of each separat wheel. Systems[18] like this are used for increasing driving dynamics.

[11] unsettling of the driver
[12] targeted steering in oncoming traffic
[13] provocation accidents appearing accidentally
[14] combustion engine or electric drive
[15] electro-hydraulic clutches (e.g. Haldex)
[16] high power and torque, also high drag torque (motor braking)
[17] torque reduction, engine speed adjustments
[18] e.g. Audi quattro with sports differential, BMW xDrive

Objective Powertrain

Regardless of whether an attack on the drive, the transmission is carried out or the 4wheel drive, the resulting symptoms are almost identical.

Changes of the drive torque[19] lead to either driving instabilities or damage resulting from the vehicle rest.

The physics behind the driving stability are quite complex. So, detailed attacks would need a lot of knowledge about the behaviour of those systems.

Attacks in this way cannot be excluded.

[19] especially to a single wheel

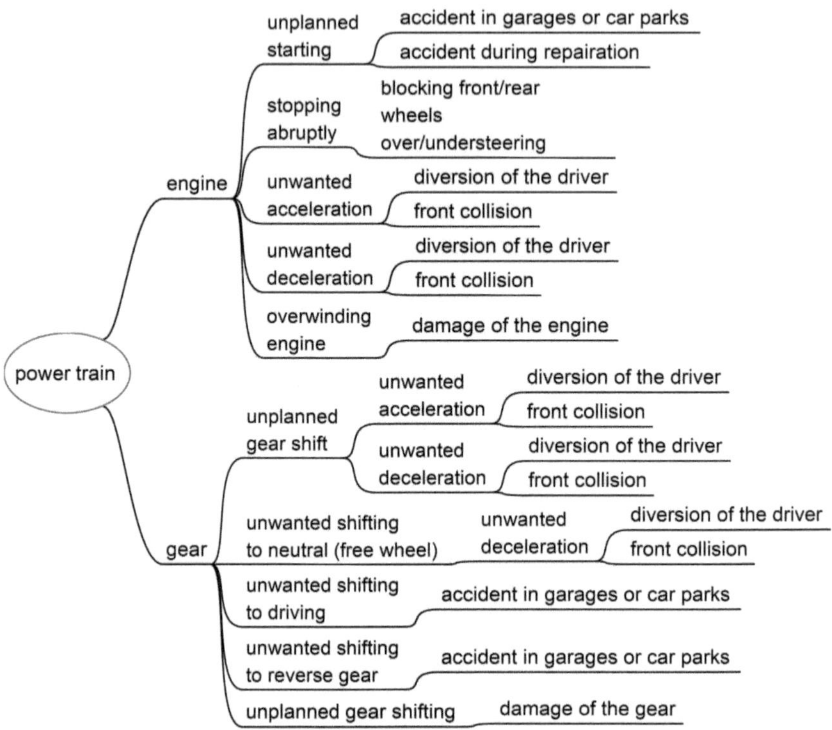

Fig. 31.: impact powertrain

4.2 Perception

In this chapter, it is no longer about physical attacks and interventions, but the perception, which perhaps sounds a little bit unusual at first glance.

The unexpected and unsuspected aspects offer a variety of subtle interventions, which should be very difficult to understand and reconstruct.

Therefore, the credibility of witnesses[20] needs to be reassessed[21]. Criminologists have to consider completely new technical possibilities.

4.2.1 Eyesight Effects

Sight is probably the most important sensory perception in traffic and driving.

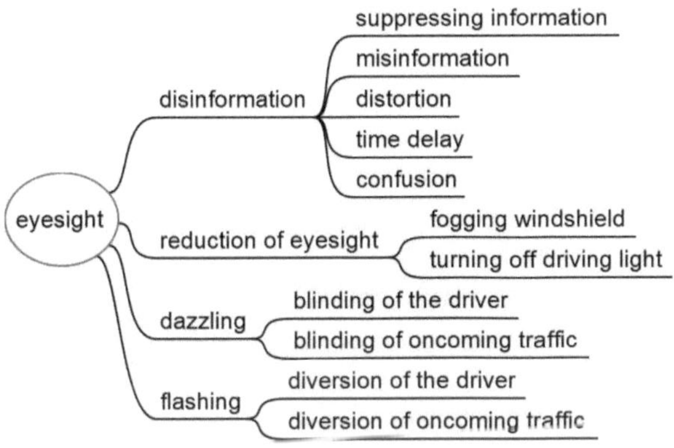

Fig. 32.: impact eyesight

Especially if one sense for environmental perception and information retrieval is preferred so much, attacks on it are very effective.

[20] driver, passenger

[21] strange and unbelievable experiences of the witnesses could be very true

Objective Disinformation

Already in normal road traffic the driver is faced with a flood of information. Information and warning signs are only a small percentage here. To assist the driver, there are an increasing number of systems that detect traffic signs on camera systems and put on different displays available to the driver.

If a driver rely on this information, probably in combination with modified vehicle speed, manipulations are at least a good prank, e.g. suggesting higher speed limits. Speed traps are welcome.

Fig. 33.: Mercedes-Benz C-Class head up display [29]
photo by Mercedes-Benz

Not only the environment provides information. Also, the vehicle generates constantly information, such as status information[22] or even navigation etc. Disinformation could be a method to annoy or confuse the driver or concealment of critical status information, which contain a high destructive potential[23,24].

[22] e.g. vehicle speed, engine speed, temperatures
[23] see section 3.2 - Destructive Hacking on page 33
[24] see section 4.3 - Destruction on page 55

Visual information are provided by camera based systems, too, such as rear view cameras to put right unclear vehicle dimensions.

Systems like this offer a lot of possibilities for attacks. The whole signal path can be influenced. Starting with modifications of camera internal image correction to solve image defects such as lens distortion or brightness adjustments. Even small changes can complicate the interpretation[25] of the image or make impossible[26].

Fig. 34.: Panasonic rearview camera module [30]
photo by Panasonic

In the signal transmission interesting manipulations are also conceivable. A simple time delay can cause a significant misjudgment, e.g. when overtaking.

Manipulated overlay graphics can make the camera image useless, cleverly changed helping lines[27] can cause accidents, too.

Attacks on the display can have various effects, such as confusing[28], dazzling[29] and flashing[30].

[25] distance estimation by distortions
[26] miscoloring or loss of contrasts
[27] e.g. rear view systems, Figure 9
[28] mirroring screen content
[29] white and white screen content
[30] inserting flashing white frames in the video stream

Objective Reduction of Eyesight

A very easy way to affect the visibility is to reduce the sight of the driver. Impossible? Not really! All it needs is one flap in the heating system closed and the windows steam up very quickly.

Also very effective could be turning off the light when it is dark[31].

These methods are just individual interventions. In combination with other operations the hacking effects can be increased easily.

Fig. 35.: limited view at night
photo by author

[31] very interesting in combination with measured light situation by common rain-light-sensors

Objective Dazzling Effect

Unpredictable reactions of a bedazzled driver[32] using the installed light sources or oncoming traffic[33] especially at night or in tunnels could have a big impact on the driver's behaviour[34].

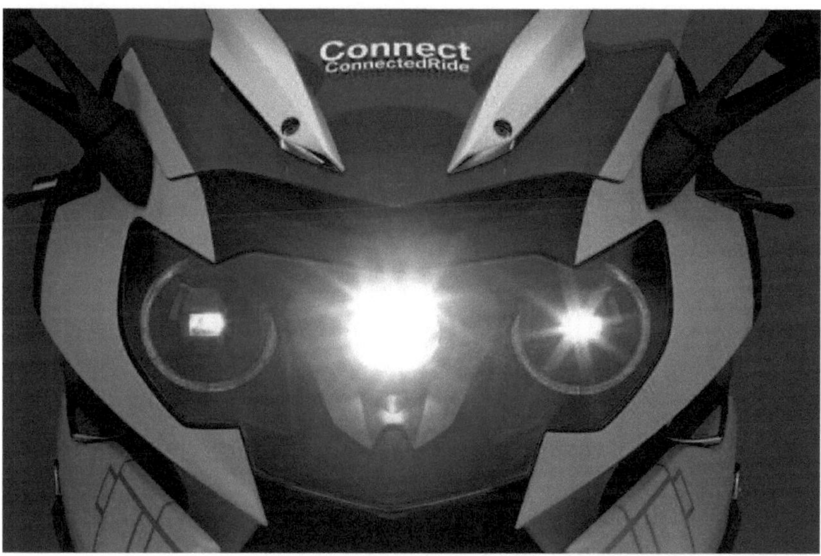

Fig. 36.: laser head lamp even for motorbike [11]
photo by BMW

Depending on the vehicle speed and intensity of the dazzling light, several seconds of blind flight could pass before a driver regains normal vision. This may be sufficient for causing an accident [52].

[32] too bright instrument clusters,
too bright displays of infotainment systems,
too bright interior illumination

[33] main beam of headlights,
headlights which produce too much light (e.g. Xenon or laser),
headlights with too small profile of light outlet (e.g. single LED),
false color of emitted light

[34] reduced control on the vehicle

Objective Flashing

Eyes can get used to constant light conditions and compensate for both deficiency[35] and excessive[36] brightness. Problematic are short changes of light conditions like flashes or warning lights.

Depending on the intensity of lights flashing or cyclical illumination the driver[37] or other road users[38] can be distracted.

4.2.2 Acoustic Effects

The impact of acoustic effects for the driver seems to be much less important than eyesight, but one is subject to a deception here.

Audio effects can have an extremely subtle impact on psyche and body. The effects [48] need not even be consciously perceived.

Objective Annoying Noises

How unpleasent and annoying a rattle and crackle in the car is, should be readily comprehensive.

What if these noises are caused by attacks via electronics? There are more than enough options to create noise in modern cars electronically. Loudspeakers[39] are the obvious means.

However, other actuators[40] may cause several noises, too. The ways to confuse and distract the driver are manifold.

While squeaking, rattling and roaring merely cause annoyance, low-frequency sounds[41] can even cause cardiac arrhythmia [46].

[35] see Figure 4.2.1 - Objective Reduction of Eyesight on page 48

[36] see Figure 4.2.1 - Objective Dazzling Effect on page 49

[37] interior lights, warning lamps, backlights of instrument cluster or infotainment system displays

[38] headlights, rear lights, indicator lights

[39] infotainment systems, instrument cluster (indicator clicks), exhaust sound actuator (sound design)

[40] flaps, seat adjustment motors, fans, pumps etc.

[41] inaudible infrasound

Objective Drown out Noises or Noise Cancellation

The selective overtones or cancelling of traffic or machinery noise is much more subtle and hard to identify in case of a hacking caused accident.

Covering or cancelling a defective caused rattling can lead to damage to the vehicle up to the accident.

Further applications are only a matter of destructive imagination.

Objective Acoustic Shock

Shocking experiences when the music all of a sudden becomes too loud is a circumstance every driver should be aware of.

In case of music noises are not really frightening. What if nonlocalizable screaming resounds in the car at once? What about virtual gunshots sounding in a 5.1 surround sound systems available in many cars? What about sounds with psychoacoustic effects? Sounds like these can frighten or irritate a driver easily.

There is a wide variety of possible sounds and acoustic shocks that can be used perfidiously.

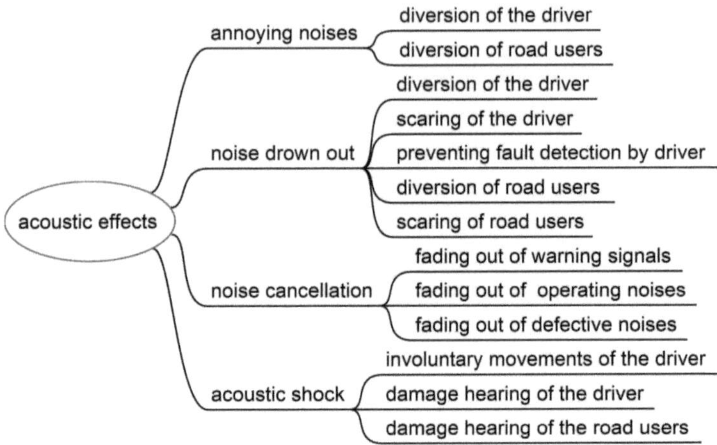

Fig. 37.: impact acoustic effects

4.2.3 Sense of Touch

At first glance the physical tacticle feeling seems to be without an effect in the vehicle.

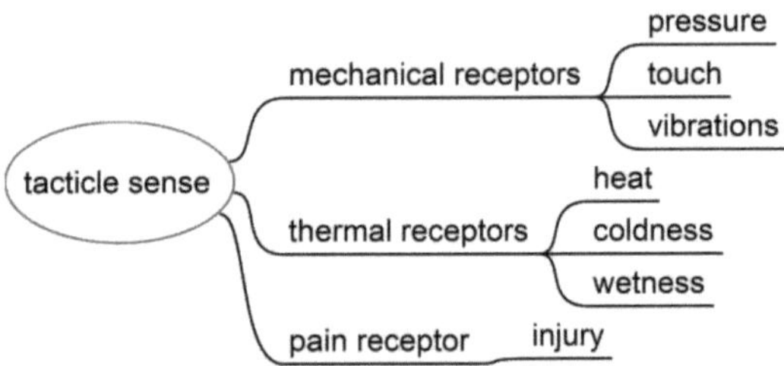

Fig. 38.: tactical senses

However, this sense of the driver has much more impact on his driving behaviour and wellbeing as expected.

The contact points between the body of the driver and the vehicle are very diverse and highly sensitive. Accordingly manipulations are conceivable and feasible.

Objective Sense of Touch

A vibrating steering wheel, an over tempered seat heating, an overheated cabin, an unexpected cold breeze in the face can have an impact on the concentration of the driver or can distract him.

Attacks on the body's sense can be done very subtly and cause nevertheless in combination with other attacks a lot of damage.

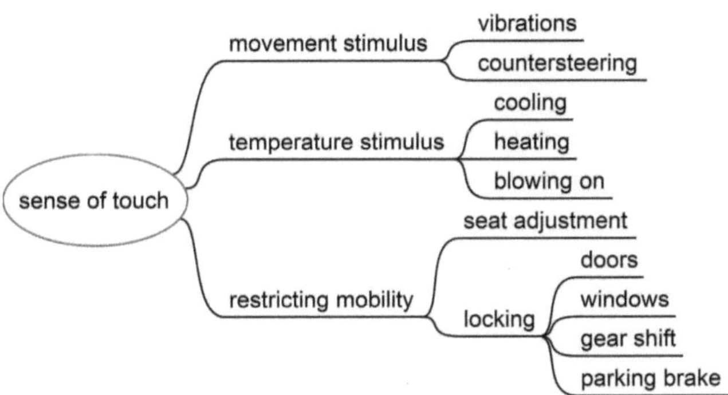

Fig. 39.: impacts sense of touch

4.2.4 Distraction

What a wonderful world of possibilities are distractions. It seems to be obsolete to add further distractions by hacking, there are more than enough implemented yet in modern cars by the manufacturer, so why adding new?!

That's a legitimate question. Here is a short overview about some implemented systems

- too many buttons in dashboard[42]
- overloaded instrument cluster[43]
- too many textual warning messages on different screens[44]
- overloaded infotainment systems[45]
- overloaded navigation systems[46]
- too many animations on infotainment displays[47]

- highly configurable vehicle settings[42]
- highly configurable head up displays
- highly adjustable electric seats[42]
- too many warning sounds[48]
- linked smart phones[49]

What about the smart phone in the driver's hand used illegaly? What about the smart watch on a driver's wrist?

Do additional distractions by hackers really make sense or are the implemented controls, displays and gadgetries more than enough?

A closer look at the motivations of hackers[50] gives the answer: Yes. Additional sounds and further text messages can mislead[51] or disturb the concentration of the driver vastely.

In fact, additional distraction by hacks may only be a drop in the ocean, but they should not be neglected.

4.2.5 Summary - Perception

Attacks on the perception of the driver and passengers can be either extremely direct effect[52], but are also used extremely subtle[53] or long lasting[54].

---

[42] play instinct

[43] information overload

[44] confusion

[45] interactive manual, eco drive graphics and animations, hybrid monitors

[46] travel guides, points of interests, best routes

[47] air-condition animation despite separate air-condition control unit

[48] belt warner, park distance control, ignition on, parking brake warning, tyre pressure monitoring etc.

[49] Internet radio, youtube, further media sources

[50] see chapter 3 - Motivation of Hacking on page 30

[51] e.g. manipulated navigation system or traffic messages

[52] e.g. blinding, acoustic shock

[53] e.g. noise cancelling

[54] psychological effects

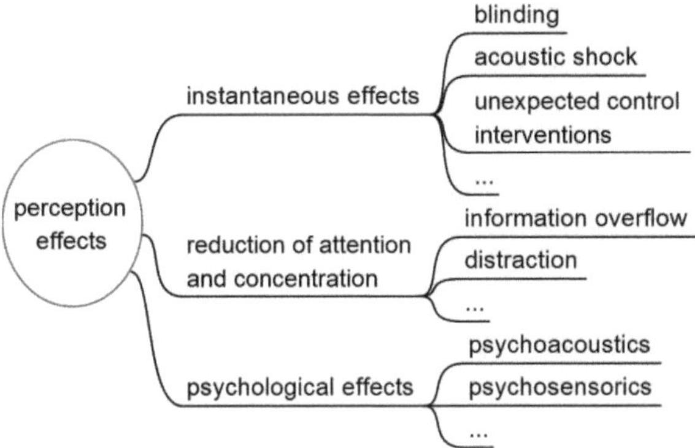

Fig. 40.: perception effects

A reconsideration of really long established systems[55] with impact on the human body and its senses is therefore recommended.

This means additional topics for psychologists already employed in the automotive industry, too.

4.3 Destruction

In current vehicles there are a lot of actuators in use. There are electric motors for fuel pumps, fans, window lifter, magnetic valves, servomotors for flaps, seat adjustments, several heaters, lamps[56], Peltier elements[57], speakers[58] and many more.

Are these actuators not a direct invitation for abuse by hackers? Their possible options seems to be almost endless.

[55] e.g. multimedia systems, illumination, cooling, motorized components

[56] interior and exterior illuination, mostly LED nowadays

[57] coolable cup holders

[58] car HiFi, instrument cluster, exhaust sound actuators

At least the wide field of network communication[59] and applied diagnostics[60] is a helpful toolset[61,62] ready for use by hackers.

4.3.1 Overstraining

All these actuators were developed and optimised[63] for their specific application. But what if they are overloaded? What, for instance, if they have to drive much more cycles as planned?

Overcurrent

Actuators are often designed with performance reserves. Fine tuning to the necessary, material-saving power consumption is realised by controls in many cases.

In most cases overstraining by power consumption is prevented by fuses, but sometimes not, especially for reasons functional safety or necessary high availability.

Only detailed analyses and measures can prevent destroyed actuators or cable fire.

Cyclic Overstraining

Quite interesting for hacking might be overstraining by cycles instead of a single, destroying event.

Manufacturers usually plan a certain number of cycles for actuators based on life time of the vehicle. More cycles accordingly lead to increased abrasion and finally to destruction[64].

---

[59] see chapter 10 - Network Communication on page 109

[60] see chapter 9 - Applied Diagnostics on page 84

[61] see subsection 9.3.1 - Selective Actuator Tests on page 97

[62] see subsection 9.3.2 - Actuator Tests by Sequence on page 100

[63] strain profile, power consumption, weight, costs etc.

[64] warranty costs, annoyed customers

Waveforms

This method certainly requires a lot of detailed knowledge in electrical engineering, driving actuators and signal analysis.

Some actuators[65] are driven by electrical current in special waveforms[66].

Manipulations[67] of the power electronics would have no effect - at first glance. In the most favorable case the harmonics[68] of those manipulated waveforms only lead to noises[69] that is difficult to identify.

In the worst case, however, an increased abrasion of mechanical parts[70] due to vibrations.

4.3.2 Movements, Limit Stops

Most applications require a complex interplay of various actuators and mechanical parts. Dependencies between positions of actuators are very typical.

At least an attack on this interplay could disturb the system interactions quite easily. This method would be brutal but effective.

Oversteering over given limits, can also lead to the destruction of the actuator and involved components.

4.3.3 Timeouts, Violation of Time Limits

Some actuators[71] can overheat easily. Therefore, time limits are defined by supplier of the actuators or actuating elements. Particularly

[65] especially electric motors

[66] e.g. pulse width modulation, sinus waves, square waves, triangle wave

[67] e.g. sinus wave to square wave

[68] multiples of the base frequency

[69] see subsection 4.2.2 - Objective Annoying Noises on page 50

[70] e.g. (hairline) cracks

[71] e.g. magnetic valves, magnetic dampers

discrete[72] actuators[73] are usually completely unprotected against overheating.

Protection against overheating is mostly implemented via software, Hardware solutions[74] would be too expensive in many cases. Therefore, overstraining by overheating is an option that is quite easy to understand.

As already mentioned, for hackers there are several tools[75,76] available for this purpose.

---

[72]actuators without microcontrollers

[73]e.g. window or mirror defroster, windshield washer pumps, radiator fans

[74]thermal fuses

[75]see chapter 9 - Applied Diagnostics on page 84

[76]see chapter 10 - Network Communication on page 109

5 Situational Hacking

An important question when dealing with hacks is the moment and duration of an attack. Therefore, this chapter deals with the question of **WHEN** a hack occurs and **HOW LONG** it lasts.

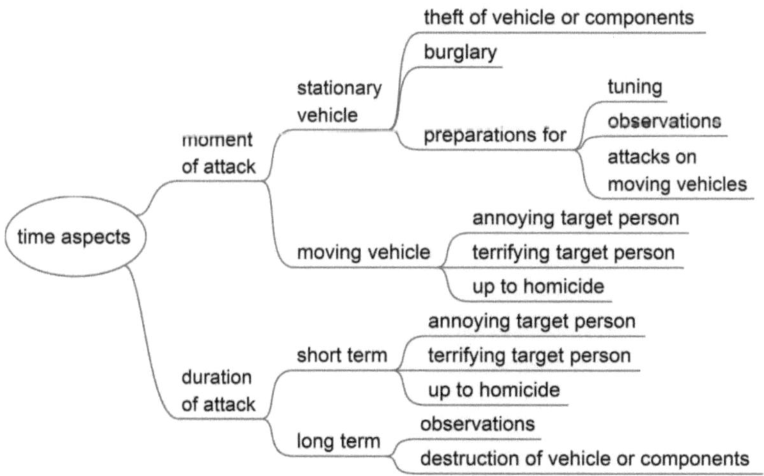

Fig. 41.: attack scenarios by time aspects

5.1 Moment of the Attack

If the moment of an attack could not be accurately determined, the driver would have the chance to be prepared and it would be easier to fight off attacks in some cases. But there are various possibilities and attack scenarios conceivable, especially with a criminal intent.

Here, among others, counts the element of surprise, the actual situation[1] and the duration of action[2], that can make a difference between annoying and killing of a target person.

Tab. 5.1.: Moments of attacks

Available Interfaces	**Parking, locked car**	**Parking, un-locked car**	**workshop**	**Red traffic lights**	**While driving**
GSM	x	x	x	x	x
Bluetooth			x	x	x
Bluetooth OBD adaptor				x	x
OBD plug		x	x		
remote key receiver	x	x	x	x	x
vehicle network		x	x		
Wi-Fi	x	x	x	x	x

Another situational factor is also the accessibility of interfaces that can be used for hacking attacks.

[1] parking the car or driving full speed on the highway

[2] third-party intervention by short braking or permanent emergency braking, short steering wheel twitching or turning maneuver

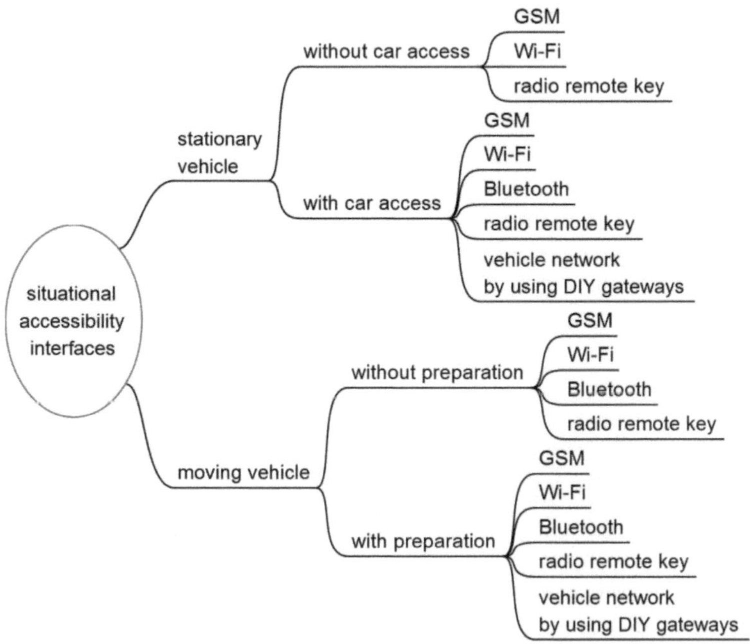

Fig. 42.: accessibility of interfaces by situation

5.1.1 Stationary Vehicle

A stationary vehicle[3] seems to be less attractive for hacking attacks. But the impression is deceptive.

In many cases it gives the chance to create the basic requirements for hacking - this is access to vehicle electronics.

Parking Without Car Access

A typical situation for a stationary vehicle is a locked parking[4] vehicle.

The number of possible attacks on cars in sleep mode is not that high, because to protect the battery most of the electronic systems are

[3]vehicle with zero speed

[4]car in sleep mode

switched[5] off. Exceptions are radio receivers for remote keys, GSM modules[6] or satellite tracking systems[7].

Fig. 43.: The number of keyless car thefts is rising [32]

In some cases "convenient" car lock systems allow a quite intelligent burglery or theft [32]. So called keyless or passive entry systems can be tricked by extending the radio range of the keys[8] via radio sets [53] working as repeater.

A special use case in the future will be the electric vehicle at the charging station. In charging mode, certain number of control units[9] cannot be turned off, because the charging of batteries on the vehicle requires a lot of communication[10].

[5] sleep mode, wake up concepts

[6] used for anti theft systems or diagnostic topics (firmware update over the air)

[7] GPS based

[8] about 1 m range between remote key and vehicle

[9] HV battery ECU, on board charger ECU

[10] billing information, state of charge, health of battery

Especially in case of wireless charging Wi-Fi modules will be running all the time during the charging process and have Internet access so as to communicate the battery status to the driver via smart phone.

First targets for attacks could be the gateways. Therefore, gateways are developed with special attention and confidentiality.

Parking with Car Access, Garage

Forgetting to lock the car can easily provide additional interfaces for hacking easily. Using tiny hardware units[11] an access to the whole vehicle network would be made available.

The fact remains that a DIY gateway to unlock the TV feature of a vehicle infotainment system has been spotted several years ago.

Access to the vehicle network opens up many new options for "good" and "bad" hackers.

5.1.2 Vehicle in Motion

On the road the whole vehicle with all its systems and functions is active and can be used for hacks nearly without limits in case of vehicle network access. This means, that for instance Bluetooth and Wi-Fi connections[12] are waiting to be hacked.

Nobody should rest just by guessing that the Bluetooth protocol is unhackable. It is not. There are lots of tools for Bluetooth and Wi-Fi hacking available in the depths of the Internet.

An invader sitting in the next car could attack in a very unobtrusive way and would be able to observe the hacking consequences at the same time.

[11] hacking gateways

[12] infotainment systems, Bluetooth OBD dongles

Fig. 44.: vehicle in motion
photo by author

A hack using the GSM units would have more impact and suprise effect. The possibilities themselves for hacking of wireless data connections to the vehicle do not differ.

5.2 Duration of the Attack

An important parameter for attacks on the vehicle is the situation.

Timing and duration determine the effect of the attack on man and material or theft[13] impacts. It is easy to understand that abrupt interventions have a very direct impact on man and material.

Considering the duration of the attacks, there are significant differences in the attack targets. Short-lasting interventions are likely to have the greatest possible extent geared to theft or direct and harsh attacks on life and limb.

In contrast, long term interferences are much more directed to observations, destructive or annoying attacks, see Figure 41 on page 59.

[13]vehicle, objects or data

6 Summary - Hacking Options

The good news at first: Most of these hacking options are more or less theoretical.

Tab. 6.1.: Short overview of hacking options

		attack		impact		mode of action							type of damage		
							cognitive impact								
target system	loss of controllability	spontaneous	insidious	non-destructive	destructive	driving stability	eyesight effects	hearing effects	tactile effects	diversion	scaring	overload	just annoying	material, vehicle	live & limp
4wheel drive	xx	x	x		x	x			x		x	x		x	x
automated gears	xxx	x			x	x						x		x	x
battery[1] systems	-		x		x								x	x	
braking systems	xxx	x			x	x			x		x	x		x	x
car security[2]	-	x			x									x	
air condition controls[3]	x		x		x		x			x					x
damping control	xx		x		x	x			x		x	x		x	x
door modules[4]	x	x		x						x	x		x		x
HV battery systems	xxx		x		x							x		x	x
instrument cluster	x	x		x			x	x		x			x		x
light control[5]	x	x		x			x			x	x		x		x
infotainment systems	x	x	x	x	x		x	x		x	x	x	x		x
safety modules[6]	xxx	x			x		x				x			x	x
seat adjustment modules	-			x					x	x	x	x	x	x	
steering systems	xxx	x			x	x			x	x	x	x		x	x

This does not mean that those options will not be used in the future. Therefore, they need to be the object of appropriate measures.

[1] 12/24 V

[2] remote keys

[3] separat unit for air condition system

[4] lock, window lifter

[5] body domain controller

[6] airbag, pretensioners

Some fundamental considerations have to be made. An assistance for this purpose shall be given in Figure 46, too.

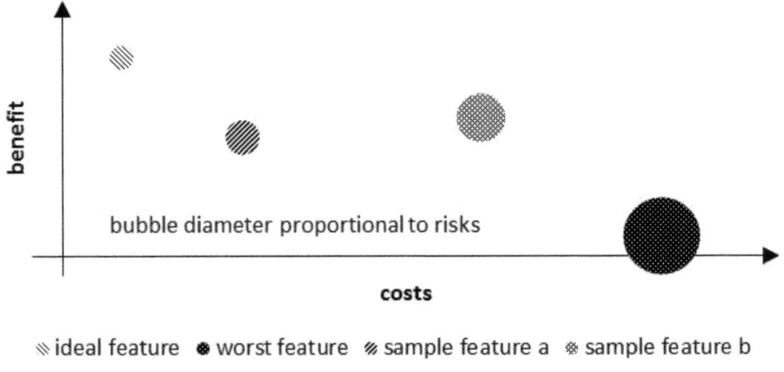

Fig. 45.: rating of benefit, costs and risks

The main questions that have to be answered are:

- Is the relationship between benefit and risks of the feature really analysed adequately?
- Are the costs of the feature fully covered?
- Is this feature sufficiently analysed for customer acceptance?
- Does the feature include the risk of reputational damage?

A diagram showing the relationships as seen in Figure 45 may be helpful for finding an entire decision. However, it presupposes an intelligent way of quantifying and certain honesty in rating of the benefits.

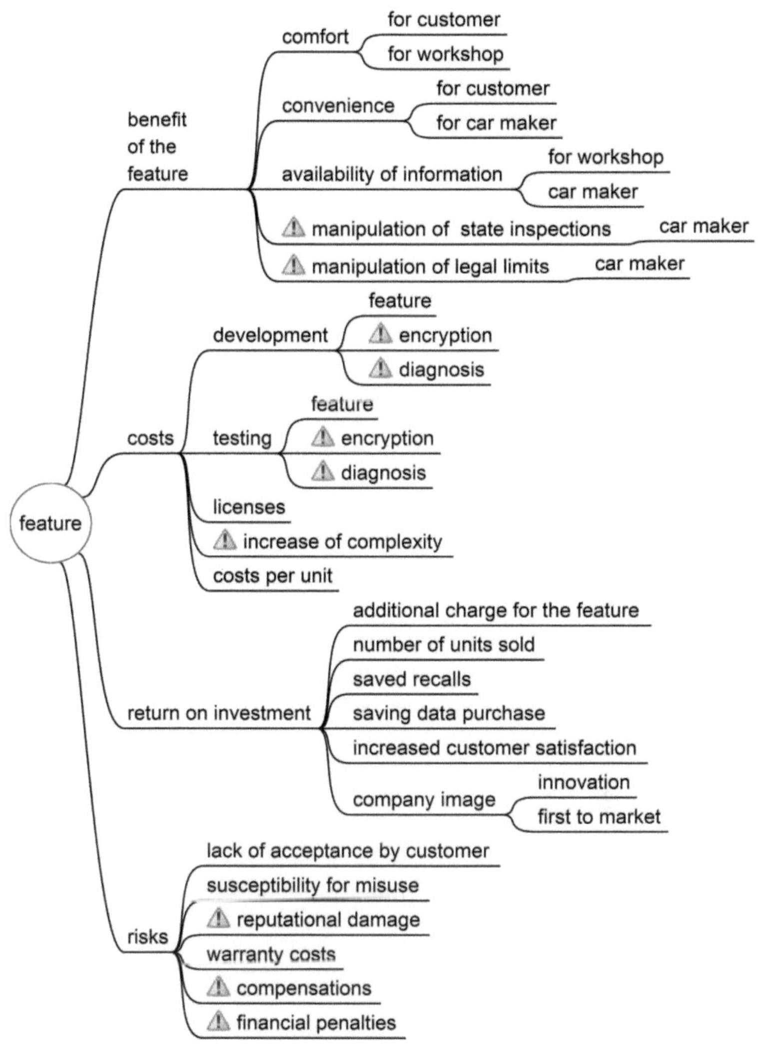

Fig. 46.: decision finding aspects of features (critical or underestimated aspects are marked)

Part III.

Methods of Hacking

After the questions about the *WHO*, *WHY*, *WHEN* and *WHAT* have been clarified, there is now the question of **HOW**.

While the recent topics largely were still quite explainable without special background knowledge, it goes into more detail now.

Therefore, some topics of the vehicle electronics should be subject to deeper contemplation.

Again, of course, this book shall not be read as an instructions for hackers. It will rather give a sense of where potential technical weaknesses lie.

7 Features and Functions

The basis for the understanding of the electronic components is a concept of features and functions in all its aspects. A special difficulty is the perception of the function by a lack of differentiation towards the feature.

Therefore, initially the wording of a few technical contexts needs to be clarified.

7.1 Customer Usable Function

At the first glance customer usable functions are quite easy to explain. But in fact it is not that easy as it looks.

Not only the various manufacturers, but also company divisions and departments have partly fundamentally different views.

The languages in use generate variants of meanings of terms, too. Not only the level of proficiency[1] but also the language itself[2] plays a major role.

7.1.1 Identification Customers

Already in the question of the customers there are different opinions depending on the viewer[3]. To think of the driver only as a customer is not longer enough. An extension of the user's circle based on stakeholder would fit much better, see Figure 47.

[1] foreign language vs. native tongue

[2] missing or extra meanings of terms

[3] R&D, sales, production line, customer service etc.

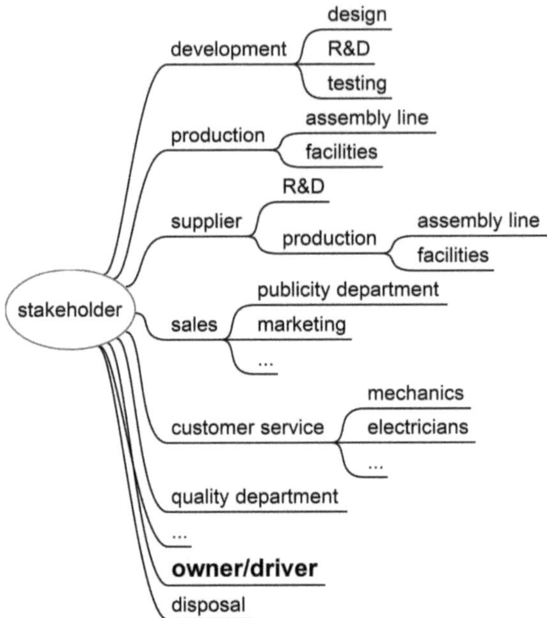

Fig. 47.: rough outline of the stakeholders

Looking at the driver with all facets, it is easy to find out that customer satisfaction is not only based on design, vehicle behaviour and a filled feature list.

A predominant share of the profits by automobiles comes from the customer service. The expectations of the customer "*driver*" for fast, inexpensive service and targeted repairs is quite understandable.

Ultimately, the following customers and their particular goals should be considered in all deliberations.

Including the garages with customer service in the group of customers is quite unusual in thinking of current developments, but neccessary to get a well designed product vehicle in all relevant use cases.

Tab. 7.1.: Overview customer and goals

	comfort	driving assistence	infotainment	navigation	reparability	safety functions	security functions	USP
dealer					x	x	x	x
driver	x	x	x	x		x		
owner					x		x	
passenger	x		x			x		
workshop					x	x		

7.1.2 Typical Function Set

Especially in luxury vehicles, there is an almost overwhelming number of features experienced by the customer.

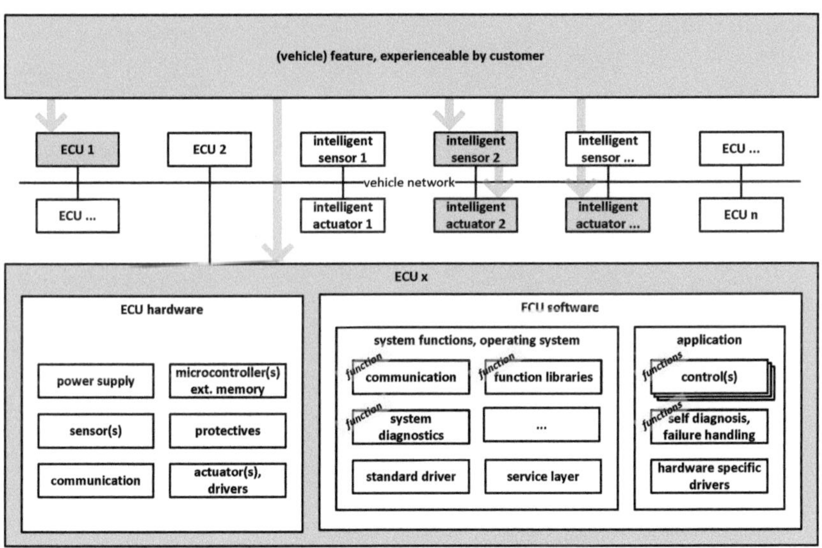

Fig. 48.: feature vs. function (I)

This results in an even greater number of individual functions, which are necessary for realizing the features. This connection is shown in Figure 48.

In Figure 49 there is a trial systematisation of relationships and a breakdown in the typical fields of customer usable features and system functions.

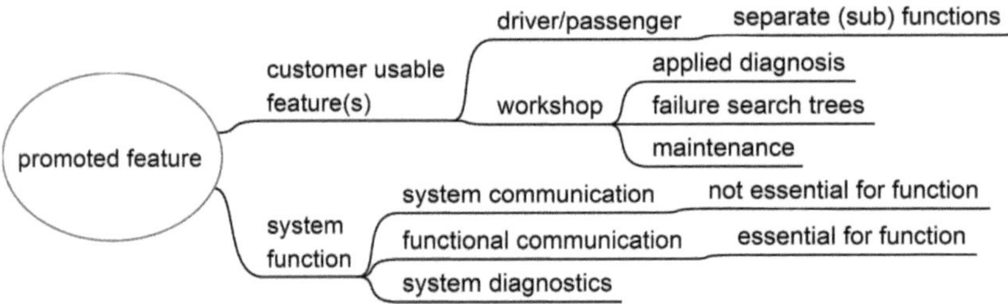

Fig. 49.: feature vs. function (II)

What does this mean in practice? The best way to illustrate it is by using an example.

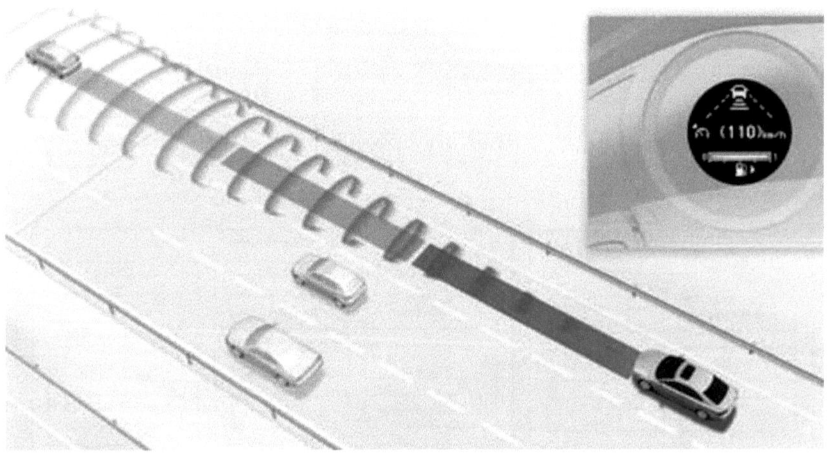

Fig. 50.: Volvo Adaptive Cruise Control illustration [1]

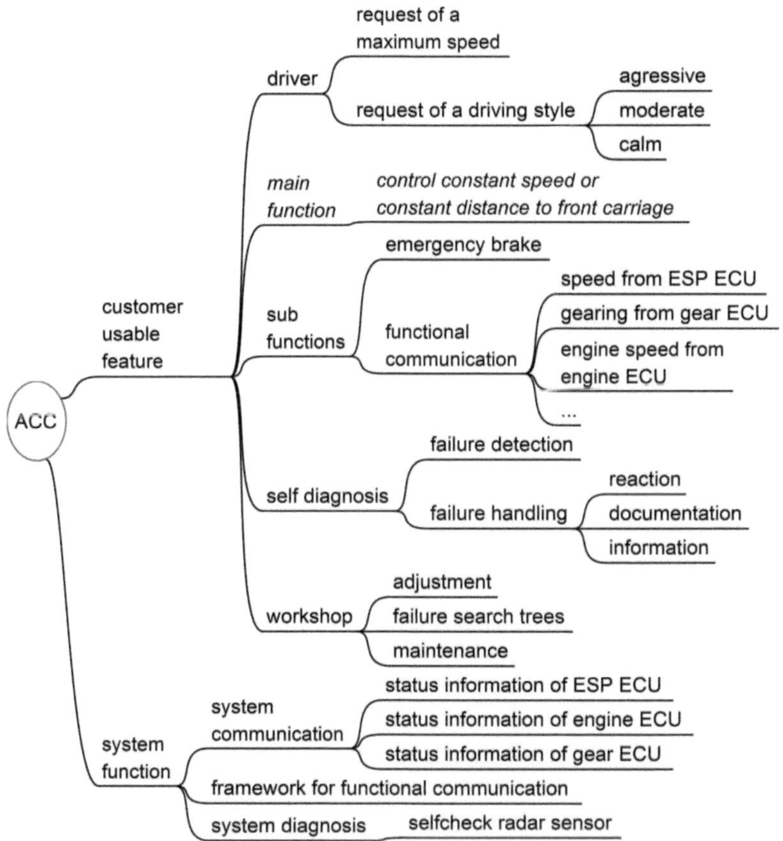

Fig. 51.: functions of feature "Adaptive Cruise Control"

In order to fulfill these tasks, various ECUs must be networked and collaborate.

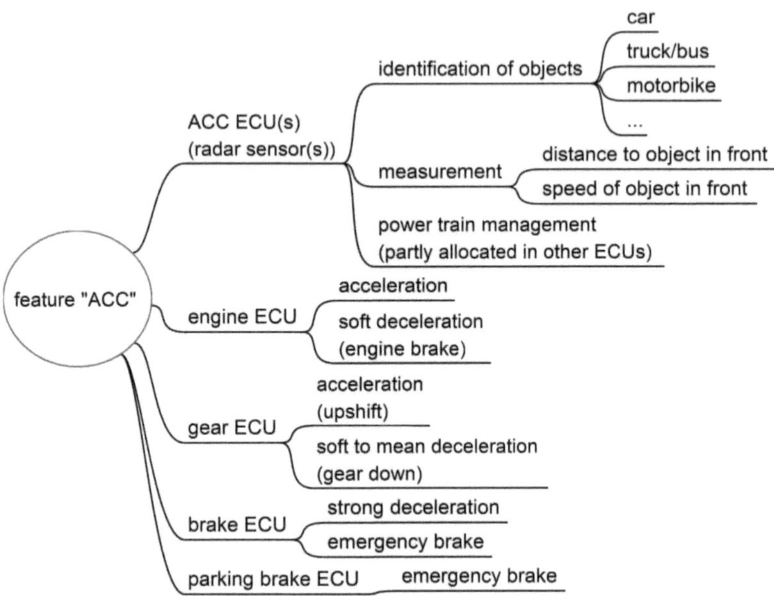

Fig. 52.: required ECUs for feature "Adaptive Cruise Control"

In effect, only a single function for the driver will be experienced. The importance of the hacking is shown in greater detail in chapter 8 - Controller on page 77.

Most of the other functions that are necessary for the functional implementation of the features are more or less completely hidden from the driver, such as self-diagnostics[4], workshop diagnostics, functional communication etc.

7.2 System Functions

Until now the example of Figure 51 was considered in question of functionality. The focus was on functions, that were obviously recognizable.

[4]failure detection, failure handling

This section is about less obvious parts of the functions, software parts which are essential for for the correct implementation of features in the vehicle system.

7.2.1 Communication Framework

A large part of the system functions is covered by the communication. The communication framework includes tasks such as addressing, timing, multiplexing, signal handling, end-to-end protection, signal plausibility[5], failure handling[6] etc.

The communication framework contains hardware-related configurations[7], start up behaviour, signal plausibility checks, sleep modes and power management, too.

More detailed information about communication see chapter 10 - Network Communication on page 109.

7.2.2 Diagnostics Framework

Also very extensive is the framework that enables the diagnostics of communication to the outside[8].

It includes the necessary protocols including timings, handshake, methods and parameters.

7.2.3 Operating System

Both the communication and diagnostics are based on standardised rules. Other system functions are also largely similar if not identical.

Thus, it makes sense that functions which can be reused would be combined in a modular operating system.

In most cases, control units are equipped with operating systems that are based on the AUTOSAR standard.

---

[5]range checks

[6]time outs, incorrect frame counters, cyclic redundancy checks (CRC)

[7]transceiver configurations

[8]offboard communication

The advantages are obvious. Bugs are found quickly through the wide use. Also, errors that were not discovered in a particular control unit but in other systems can thus be fixed.

7.3 Self-Diagnostics

Self-diagnostics is a quite complex topic with various perspectives. As a part of the functionalities the self-diagnostics has mainly to be seen as failure detection and information provider.

The various functionalities of self-diagnostics are part of the applied diagnostics and treated in subsection 9.1.2 - On-Board diagnostics on page 87.

8 Controller

Many functions in the vehicle have a direct reaction with the physical environment. In a simple case a temperature or a velocity is kept constant.

Already behind these seemingly trivial controls, quite complex technical interactions in hardware and software can be found.

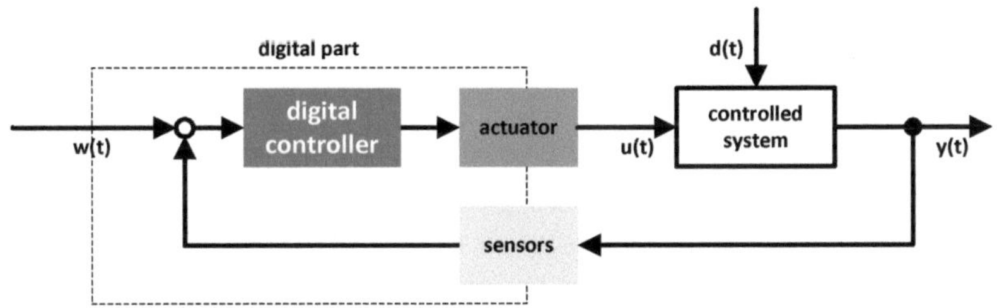

Fig. 53.: closed loop control [51]

A brief overview of the relationships of a control in a closed loop is shown in Figure 53, a very simplified scheme.

However, the relevant parts for attacking possibilities are shown.

8.1 Sensoring

The basis for controls is the measurement of the physical values which should be controlled. For this purpose, a corresponding sensor is needed.

First, sensors have to convert physical values into electrical signals, and there are many different ones. Some typical values are

- temperature
- pressure
- speed
- brightness
- acceleration
- rotation rate
- voltage
- current
- humidity
- distance
- liquid level
- limit stop

- system-rated position
- geological position
- loudness
- gas analysis

In a further step the measured values must be provided to the microcontroller available in computer readable form, digital signals[1] or analog voltages or currents[2] via ADC[3].

The versions of sensors can be very different depending on the application. A good differentiation in regard to the potential hacking is the assembly site. There are ECU internal and external sensors possible.

While internal sensors[4] usually offer little attack surface, the manipulation of external sensors is much more likely, as they are still linked via bus to the control unit.

Sensors can be manipulated in various ways. Interventions on the physical sensing[5] are possible as well as the change in physical values[6], modifications of transducers[7] or attacks via bus systems[8].

But why would a sensor ever be hacked?

Depending on the motivation of hacking[9] sensor manipulations could be used for tuning[10] as well. But also destructive[11] interventions are conceivable.

[1] intelligent sensors
[2] discrete sensors
[3] analog digital converter
[4] e.g. current sensors (shunt), pressure sensors, ECU temperature sensors
[5] e.g. reduction of sensing areas
[6] e.g. bypassing
[7] modifications of amplification
[8] e.g. offset correction or sign inversion of measured values
[9] see chapter 3 - Motivation of Hacking on page 30
[10] increase of performance from the engine
[11] see section 4.3 - Destruction on page 55

8.2 Control Algorithm

A simple description of the algorithms is hardly possible. Even for simple examples complex mathematical formulas are needed to describe the control behaviour itself.

In general, the standard algorithms[12] for controllers are not used in pure form but in various combinations of them.

The algorithm is part of the application software and typically embedded in an operating system.

Targeted changes to this programmatically converted formula works are really only possible with extremely good background knowledge. Apart from experts[13] manipulations are almost excludable.

8.3 Parameterisation of the algorithm

Although the parameterisation of controllers almost borders on magic, changes of the control parameters are quite conceivable. Of course, a lot of background knowledge is required as well. Sometimes, however, re-engineering experiments like trial and error also lead to the desired goals[14].

Hacking of parameters is facilitated by separate flashing of parameter sets[15]. A considerable effort is also required. However, this investment might be quite rewarding, if customers pay for tuning measures[16].

8.4 Actuators

Once sensors, control algorithms and their parameters have been treated, only the actuator is missing to complete the reflections about the implementation of a control.

---

[12] proportional, integral or derivative controller

[13] programmer at the system supplier

[14] for example increasing injection quantity combustion engine to increase performance

[15] see subsection 9.3.3 - Parameter Set on page 102

[16] see subsection 3.1.1 - Desired Manipulations on page 31

The actuators' purpose is to manipulate the environment in terms of the measured[17] and expected[18] physical values[19].

Since there are many different ways of influence, there are also various types of actuators and different methods of connecting to the control unit.

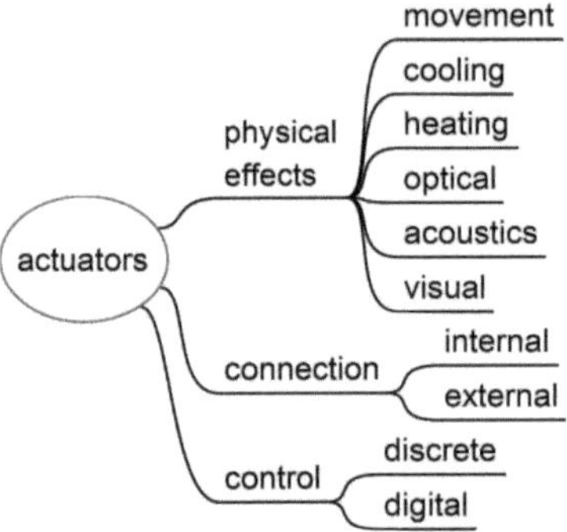

Fig. 54.: actuators principle

8.4.1 Types of Actuators

Similarly to sensors the variance of the actuators is very widely spread.

Easy to understand the distinction between physical operating principle, the mode of action.

In addition, a distinction between discrete and intelligent actuators is necessary.

Mode of Action

The types of actuators can easily be distinguished by their mode of action. A number of typical actuator types are listed in Table 8.1. Of course, the list does not cover all actuators.

[17] actual value
[18] setpoint
[19] see section 8.1 - Sensoring on page 77

Tab. 8.1.: Typical actuators

mode of action	actuator type	application
cooling	Peltier elements	cooled cup holder, cooled glove box
heating	heating wires, Peltier elements	auxiliary heating, window heater, mirror defroster, heatable cup holder
movement	electric motor, solenoids	electrical adjustments, fans, flaps, valves
visual	arc lamps, light bulbs, LED, TFT, OLED	driving and signal light, interior light, warning lamps, several displays
acoustics	speaker, piezo sounder	audio, phone, warning gongs, sound design

Discrete or Intelligent Actuators

Discrete actuators are the simplest and most financially effective form of actuators. They only contain the element for converting electrical signals into the desired physical unit. This simplicity has some advantages, but it also suffers disadvantages.

The ECU has to fulfill higher requirements, such as internal hardware driver devices, measurement devices[20], overload and EMC protections, software monitorings[21].

These requirements do not automatically have to be a disadvantage. As always, pros and cons depend on the application.

Simple actuators for single degree-of-freedom systems with few monitorings or integrated actuators with compact package can be developed in an inexpensive and effective way.

Actuators for multi-degree-of-freedom systems with several monitorings and multi-transceiver relationships are more and more implemented as intelligent actuators.

[20]several sensors for current, voltage, temperature etc.

[21]actuating variable, actuator behaviour, gradients etc.

Containing integrated logic, hardware and software driver's, measuring equipment those actuators are ECUs themselves. However, this also opens up corresponding attack opportunities that match those of pure ECUs.

8.4.2 Connecting of Actuators

Similarly, the sensors are also actuators either installed internally and controlled discretely, or running as external, intelligent actuators. The possibilities of attack are accordingly diverse.

Internal actuators are accessible to attack more or less only by software manipulation, changes in the parametrisation[22], or diagnostic procedures[23,24].

The possibilities of attacking internal actuators are only a subset of the opportunities of attacks on external, intelligent sensors exist, such as manipulations of the actuators internal sensors, actuators hardware and software or parametrisation. In addition, however, there are further options added by manipulation of network communication[25].

8.5 Summary - Controller

Finally, hacking attacks on the basis of the implemented control algorithms, sensors and actuators are quite complicated and are more likely to seek in the field of non-destructive, desired manipulations[26].

Nevertheless, even destructive intervention[27] is conceivable and feasible. However, the potential perpetrators' circle in this case is quite limited because various technical hurdles have to be taken and a lot of background knowledge must be present.

[22] see section 8.3 - Parametrisation on page 79

[23] see subsection 9.3.1 - Selective Actuator Tests on page 97

[24] see subsection 9.3.2 - Actuator Tests by Sequence on page 100

[25] see section 10.4 - Summary - Network on page 125

[26] see subsection 3.1.1 - Desired Manipulations on page 31

[27] see section 3.2 - Destructive Hacking on page 33

Evidence of changes in control units can be also brought quite[28] easily.

[28] except for complete destruction of the ECU memory

9 Applied Diagnostics

Applied diagnostics is highly dependent on the feature and corresponding applications. In practice, there is almost no control unit or function that uses all the possibilities of diagnostics methods.

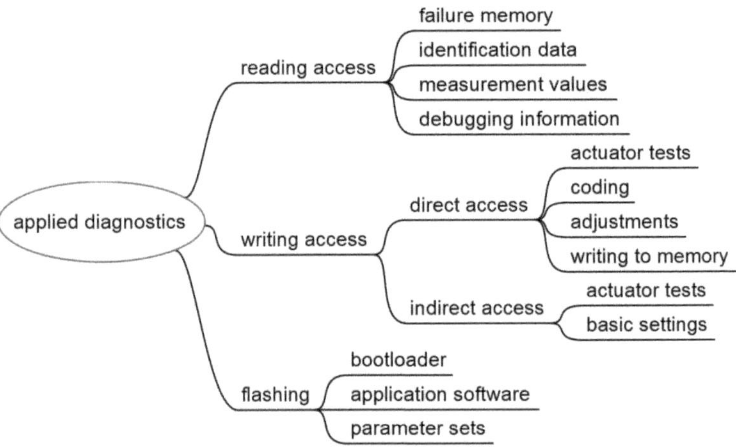

Fig. 55.: applied diagnostics

To grasp all these options requires a lot of experience if not a specialised orientation regarding the issue of diagnostics. A resulting lack of knowledge and experience about the bandwidth of diagnostic functions is quite typical and represents a non-negligible potential risk.

Not all diagnostic functions represent a potential risk. Therefore, only the main principles and methods, systematically listed in Figure 55, will be discussed here.

9.1 Short Introduction to Vehicle Diagnostics

Before embarking on the applied diagnostics, some background information should shed some light up this complex topic.

9.1.1 Users of Vehicle Diagnostics

The understanding of vehicle diagnostics varies depending on the user. In a first step, the different users and their specific views on the topic have to be identified.

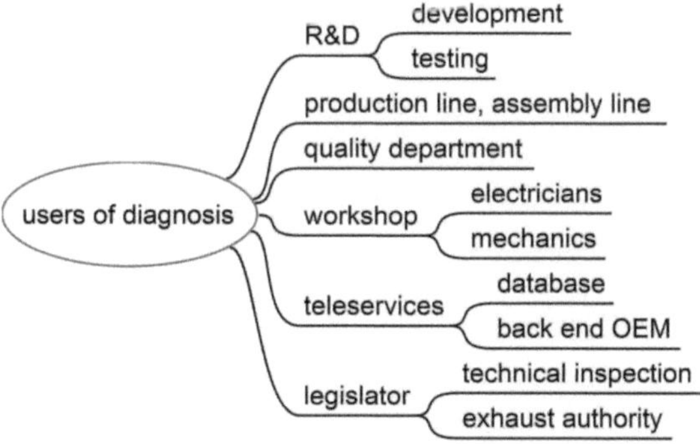

Fig. 56.: users of diagnostics

A first overview is given in Table 9.1, where the users and differences are shown. The differences in knowledge, qualification and goals listed may explain perspectives on diagnostics and the reason why it is often misunderstood.

Tab. 9.1.: typical users of diagnostics

	research and development	production line	workshop	quality department	teleservices	legislator
user	developer, tester	OEM, supplier	automotive electrician	quality engineer	telematic ECU, telemetry ECU	roadworthiness
knowledge	very specific	basic knowledge	basic knowledge	very specific	(program code)	basic knowledge, highly skilled
viewing plane	ECU, vehicle system	production specific, assembly processes[1]	vehicle[2]	ECU, vehicle[1]	vehicle[1]	vehicle[2]
qualification	engineers, engineering worker	electricians, mechanics	electricians, mechanics	engineers; engineering worker		mechanics
requried options	no restrictions on the options required	restriction on production-related options	restriction on repairs related options	no restrictions on the options required	very strict restriction on required options	restriction on examination-related options
goals	full debugging[3], parametrisation, flashing	flashing, initiation, fault-finding[4]	fault-finding[4], ECU identification, flashing	error and quality statistics	flashing, ECU status, vehicle status	evidence faultless, emissions-related fault-finding
usability	no restrictions	minimum operating options, high automation capabilities	easy to understand, automation capabilities[5]	no restrictions	no restrictions	easy to understand, automation capabilities[5]

9.1.2 Term Clarification

A very special topic in the world of automotive is the large scattering of terms and their interpretations. Each OEM is partly using special terms. In addition, the (mis)use of English and native tongue leads to further possibilities of interpretation and curiosities.

Therefore, a few basic terms have to be explained.

[1] highly OEM specific
[2] OEM indipendently
[3] up to root cause
[4] only repair relevant details
[5] guided fault-finding

On-Board diagnostics

First there is the confusing wording and concept of **on-board diagnostics** (OBD). In public, the on-board diagnostics are often reduced to the typical OBD tools[6].

But what is the real content of on-board diagnostics?

On-board diagnostics include the vehicle's own **perception**, **treatment** and **information** of failures within the ECU or function, no more and no less.

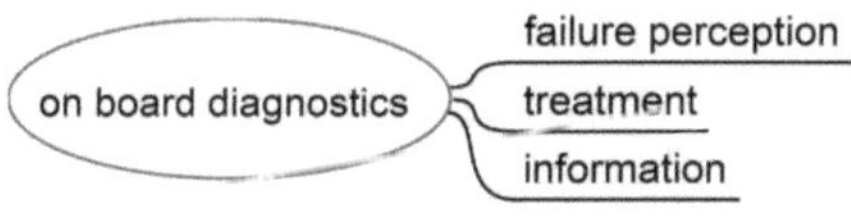

Fig. 57.: on-board diagnostics

The extent initially appears to be quite manageable. The Figure 58 shows how quickly this issue can achieve undreamed volumes.

The most important component is the detection of failures[7], the main part of the self-diagnostics. Only a detected failure can be treated. Typical failure classes are listed in Figure 58.

The range of treatments[8] can vary from switching off of a single function or complete ECU to substitution of values[9] up to neglecting[10] them.

Information is often not seen as part of the on-board diagnostics, and if, then only as a fault memory entry.

A first information about failures has to be provided to the workshop by an entry in the failure memory of the ECU or function which leads to the important term DTC - diagnostic trouble code.

---

[6] e.g. OBD Bluetooth dongle, adaptors and units for resetting service intervals

[7] perception

[8] see 11.3.1 - Responses on page 142

[9] e.g. defective sensors

[10] in case of failures without effect or minor impact

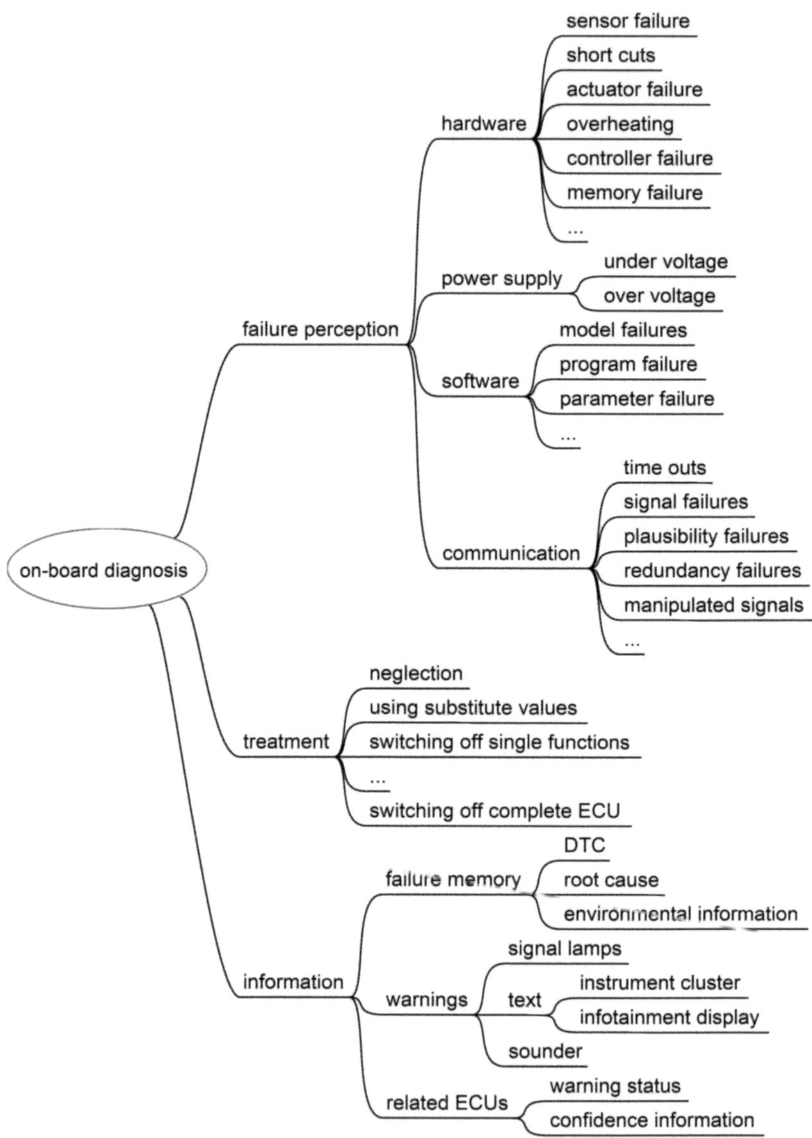

Fig. 58.: zoom into on-board diagnostics

But that's not all. In some cases, the fact that the functions the driver is used to use operate in a limited or even not at all, leads to mistakes. If malfunctions lead to a lack of driving stability[11], the driver should be informed[12] about the error.

The information could be available for the driver by warning lamps, warning gong or written text messages on the instrument cluster, too.

In a nutshell, the on-board diagnostics contain the failure detection, error handling and information on electronics and driver which are deeply integrated in the operating system of the ECU.

Off-Board Diagnostics

Contrary to the OBD, off-board diagnostics include the human-machine interfaces that do not belong to the vehicle itself. So, off-board diagnostics consist of external hardware and software.

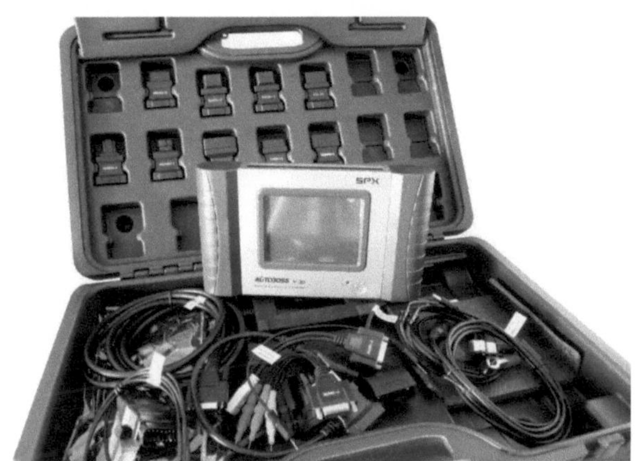

Fig. 59.: multi-brand vehicle diagnostic system
photo by Florian Schaeffer

[11]see section 4.1 - Driving Stability on page 38

[12]different perspectives of the OEMs, depending on the display strategy

Various devices are available on the market, starting from cheap and simple interface adaptor[13,14] to complete diagnostic systems for workshops[15] and production lines[16].

The available software is as manifold as the hardware. Supported are all user groups starting from do-it-yourselfer by chance to professionals as well as the different operating systems from desktop OS[17] to handheld OS[18]. These of course also distinguish the options for applied diagnostics.

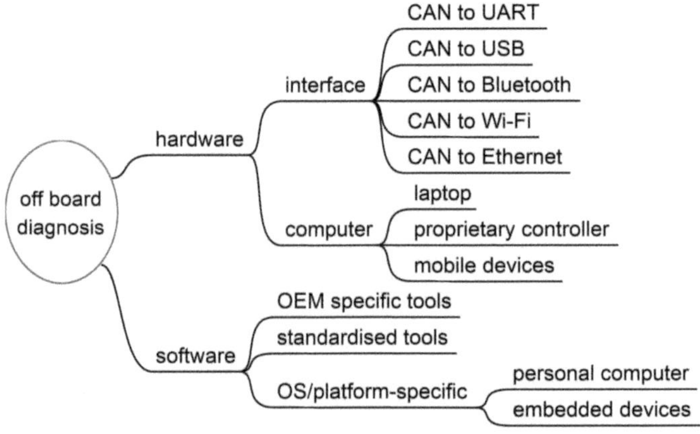

Fig. 60.: off-board diagnostics

[13] interface to several computer systems

[14] OBD plug to serial interface, USB, Ethernet, Bluetooth, Wi-Fi

[15] including different types of PC (Laptop, Tablet etc.)

[16] programmable logic controller

[17] Microsoft Windows, Mac OS, Linux

[18] Android, Apple iOS

9.2 Reading Access

Very typical and also known among the people is the option to read out data from an ECU or function[19]. After all, this is the most striking feature of applied diagnostics, but the extent of possibilities regarding the question of **what** and **how** is probably not really aware.

9.2.1 Reading Data Content

The amount of available data or information for an ECU or function could be very high. The number of the readable data depends on various factors.

- complexity of the feature and its (sub)function(s)
- number of internal[20] or external[21] sensors
- number of power supplies[22]
- number of internal[23] or external[24]s actuators
- number of connected networks
- number of monitorable values for troubleshooting

In addition to these technical factors the skills of the developer, the supplier and their experiences could enlarge the number of readable data immensely.

But which data could be read from an ECU or function?

Diagnostic trouble information

Probably the best readable value is the diagnostic trouble code. Of course, the wording as such is probably not as well known as the result - the cause of the error.

---

[19]Equating controller and function is still not common practice, but have to be taken account more and more in the future.

[20]built-in sensor in ECU

[21]directly wired (analogue) sensors or intelligent sensors connected via bus systems

[22]permanent (KL30) and/or switched supply (KL15)

[23]e.g. valves, pumps

[24]intelligent actuator

The encoded cause[25], DTC[26], priority and failure counter[27] or failure discard counter are not the only information in this context. In addition, further information[28] on error conditions is included as well.

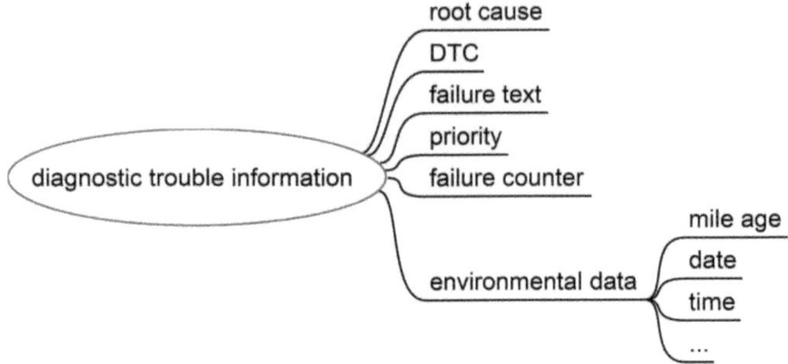

Fig. 61.: entry in failure memory

Especially the choosen environmental data that are necessary for troubleshooting is strongly dependent on the controller and its features.

Although possibly available position data or other privacy-related information can be read, they are bound to the moment of occurence of an failure. It would therefore be dependent on chance, to abouse this data.

For hacking attacks error information itself is not very productive. Seen over time, however, the request for error information may charge the controller. More see subsection 9.2.2 - Timing and Cycles on page 95.

Identification data

The amount of identification data is providing information about the diagnosed ECU or function. There is a large number of identification data containing information such as:

[25] root cause

[26] diagnostic trouble code; must not necessarily be the same as the root cause

[27] how often the error occurred

[28] environmental data (engine speed, several temperatures, battery voltages, car speed, shifted gear, etc.)

- hardware (version, part number(s)), software (version(s)), serial numbers
- flash capabilities[29]
- flashing and coding information (numbers, date, results)
- supplier of the ECU and production data[30] of the ECU

At first glance, this information is not particularly critical. If, however, the identification data would be programmed into ECUs dummies, thefts could be veiled and troubleshooting is vastly difficult.

The technical effort for developing ECU dummies just using diagnostic protocols and answering to reading requests in a proper way is not that high[31].

Measurement values

Another big block of reading data content are measurement values that may be quite interesting to hackers. Reading out measurement values is mainly used for troubleshooting and debugging.

Typical measurement values are physical values[32], status information[33], (textual) strings[34] and further counters and numbers[35].

Unlike identification data, measurement values are read cyclically and can accordingly be displayed.

In the case of measurement values significantly more opportunities for hackers are given. This is not only because of the cyclic providing of data but also the content.

[29]separate flashing of bootloader, application software and parameter sets

[30]plant number, assembly line, test stands etc.

[31]depending on used bus systems

[32]temperature, pressure, voltage, current, brightness, revolutions, vehicle speed, accelerations, rotation rates etc.

[33]on/off, failure, open/closed etc.

[34]VIN, serial numbers, names, coding information

[35]geographical position, number of passengers

Typically possible attacks do not interfere with the vehicle electronics in a direct way[36]. The criminal action is obtaining information and misuse for attacks indirectly for observation.

A small number of possible abuses in data availability are collected on Table 9.2 - Misuse measurement values for crime on page 94.

The difficulty for the developer is on the one hand to provide sufficient information by number and content for debugging, on the other hand to identify data with a potential risks and to respond accordingly. To this end, a lot of experience in troubleshooting and destructive, technical imagination is necessary.

Tab. 9.2.: Misuse measurement values for crime

measurement value	ECU or sensor	possible criminal use
distance to vehicles or objects	radar, lidar, ultrasonic	trigger for vehicle dynamic interventions
door lock status	door ECU	indicator for high jacking
exterior light brightness	rain light switch	trigger for switching off headlight, glare
humidity	clima control	trigger for windshield fogging
interior temperature	clima control	indicator for efficency of temperature attacks
level sensor	light control[37], damping systems	payload guessing
mile age	instrument cluster	selection criterion for vehicle theft
number of passengers	seat recognition	selective attacks on single persons, avoiding collateral damage
number of used key	locking system	identification of victim in case of several driver's
phone book addresses	speakerphone[38]	data theft
phone numbers	speakerphone[39], emergency call systems, telematic systems[40]	position tracking, data theft

[36]exception see subsection 9.2.2 - Timing and Cycles on page 95
[37]body domain controller
[38]infotainment systems
[39]infotainment systems
[40]diagnostics via GSM

Debugging information, memory dumps

Direct missuse of debugging information as a memory dump[41] seems to be quite inefficient.

There is simply a lot of background information and expert knowledge required to use them destructively. However, as a starting point for tuning measures[42] memory dumps can be used very well.

Like all reading access methods the timing[43] has to be taken into account. Overloading the microcontroller and the bus systems - internal[44] and external[45] - can be used for attacks, too.

9.2.2 Timing and Cycles

Reading of data via diagnostics involves not only risks by content. The timing of a reading request could have an impact in ECU behaviour, too.

Requesting a single value should not be too burdensome to the ECU. A possible exception is the reading of large data volumes[46].

In case of incorrectly configured diagnostics timing options the processor needs a lot of performance which does not cover diagnostics and objective function[47] at the same time.

Cyclic requestss or polling of measurement values could have significantly more effect on processor performance. This could not only cause performance losses for a single ECU but for an increased bus[48] load, too.

---

[41] one to one copy of data stored in random access memory or read only memory (flash ROM or EEPROM)

[42] see subsection 3.1.1 - Desired Manipulations on page 31

[43] see subsection 9.2.2 - Timing and Cycles on page 95

[44] UART, SPI, EBU etc.

[45] see section 10.1 - Automotive Bus Systems on page 110

[46] e.g. memory dumps

[47] feature experienced by customer

[48] especially CAN bus

9.2.3 Summary - Reading Access

The reading of data from a control unit via diagnostics involves a surprising number of possibilities of abuse and attacks on the vehicle as well as to life and limb.

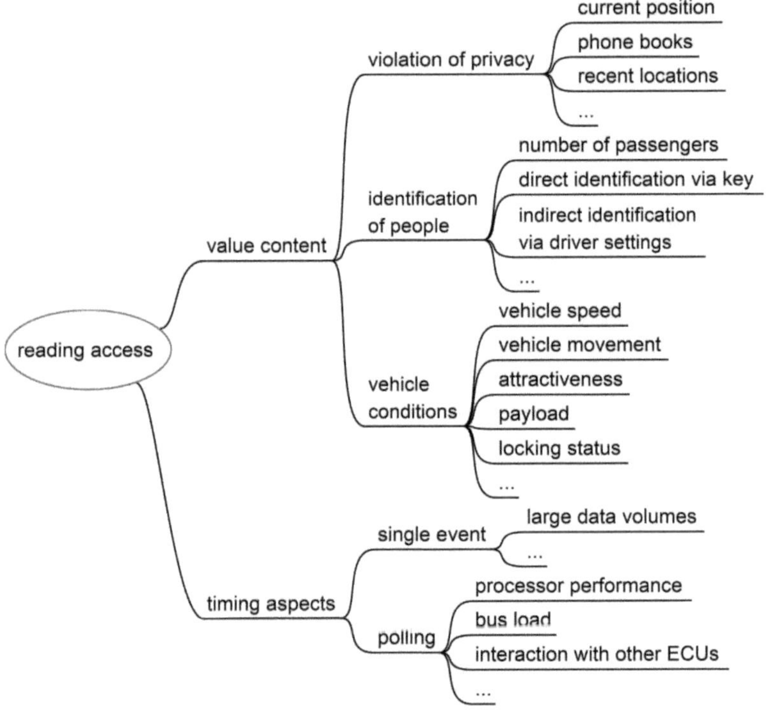

Fig. 62.: critical aspects of readings access

Therefore in the implementation of data to be read, should always be weighed the benefits and risks. Some food for thoughts about the risks are collected and systemated in Figure 62.

Finally, two important questions should be answered. **Is it possible to overload ECU and network** by reading access and **can the measured values be used for indirect attacks**[49]?

9.3 Writing Access

A lot easier to understand is the possible impact on safety and security of diagnostics by writing access. The most important methods are described below.

9.3.1 Direct Writing Access

Methods with a direct writing access can change values in the ECU directly. Depending on the method also persistent[50] but temporary[51] memory contents can be changed.

The immediacy between manipulation and system reaction makes the direct writing access for hackers particularly exciting.

Selective Actuator Tests

A very useful feature in troubleshooting is the actuator test. Actuators commanded by the ECU can be driven for testing purposes. Two different methods are available.

The first method is the selective actuator test, driving a single actuator very directly usually bypassing safeguards.

If now, for example, controlled solenoid valve manually over a longer period, it can cause overheating up to complete destruction of the component[52].

In this case, only the material would suffer. What would happen, if the steering actuator would be "tested" while driving and the vehicle would be steered into oncoming traffic?

---

[49] e.g. data theft

[50] flash memory, EEprom

[51] buffer memory for actuator driver's

[52] see section 4.3 - Destruction on page 55

In another scenario, the window lifter could be controlled by diagnostics and injure driver or passengers.

Finally, each actuator test means a hazard and should be checked according to the criteria[53,54] and implementation accordingly safety[55].

Coding

The software of ECUs is developed containing all available features and function of the control. But not all features or functions are active depending on the corresponding ECUs and purchased options.

One way to enable functions selectively is coding. The functions are assigned to bit patterns, which are then combined into one coding string and transmitted via a writing instruction to the ECU.

Using this method, the variance of ECUs, the cost of development, testing and storage can be reduced.

But this also means that not all combinations of features and functions are fully tested. In principle, this is not a disadvantage, because the logically related combinations are thoroughly tested. And in the end only these combinations can be ordered[56] by the customer.

But not only advertised functions can be switched on and off by coding. Even subfunctions can be activated and deactivated.

One attack scenario would just be to disable security features which are expected by the customer. This could lead to dramatic changes in the driving stability[57], especially if basic functions such as missing an anti-lock braking system (ABS) at once.

[53] see chapter 4 - Kinds of Manipulation on page 38
[54] see Table 6.1 - Short overview of hacking options on page 65
[55] protection by session type, security access or similar
[56] combination will be automatically checked before ordering process starts
[57] see section 4.1 - Driving Stability on page 38

Adjustments

Adjustments are a diagnostic method to the entry of numeric, vehicle-specific values.

A typical adjustment is the input of the wheel circumference in the brake control device for the precise calculation of the speed. Depending on the configuration and customer order, different wheel sizes are mounted on the vehicle ordered. Finally, the sensor of the brake control unit can detect only the wheel rotation[58].

But if incorrect values are entered now, the displayed speed may differ materially from the real speed. This is pretty annoying, especially when you are caught by a speed trap. Greetings from a hacker via speed ticket.

That would still be a comparatively harmless case, unless there are already too many collected tickets. Adjustments in other applications, however, have significantly greater impact on driving behaviour.

Therefore, the influential possibilities for hacking should be studied and possible maladjustments[59] are minimised.

Writing to Memory

Another function for debugging and troubleshooting is the direct write access to memory sections in the ECU. This method is extremely limited suitability for usage in the field. It puts ahead a very high detailed knowledge of the programming of the controller, the parameters and the memory management.

Nevertheless writes are technically possible and can be used for attacks on ECUs and vehicles.

[58] hall sensor and tone wheel

[59] plausible default values, limitation of values on plausible ranges

It is therefore important to protect these write options adequately by limiting them to corresponding sessions[60] and Security Access[61].

9.3.2 Indirect Writing Access

In some cases, vehicle-specific values of ECUs or functions must be taught, such as end stops of window lifter, level sensors in the chassis, longitudinal and lateral acceleration sensors, steering wheel sensor etc.

Here no precise data can be entered directly. Therefore routines for teaching these values are implemented and started in production line or workshop.

Although a little more detailed knowledge is required, at least manipulations are quite possible. So a misaligned steering angle sensor for instance can cause a reduced directional stability of the vehicle, a very annoying condition.

Actuator Tests by Sequence

Actuator tests have been explained in subsection 9.3.1 - Selective Actuator Tests.

Another method to drive actuator tests is by starting sequences. Contrary to selective tests a number of different actuators of the ECU are activated in a specified time sequence here.

During the testing of brake systems for instance, several valves are switched so that applying a braking force for each wheel is carried out and the correct installation of the brake lines can be tested.

For sequential tests thus result several scenarios, the activation of a sequence programmed by the manufacturer or implementation and starting of a sequence programmed by hackers.

[60] developer session

[61] diagnostic method to protect access or manipulation

Added thereto, the last type of attack is not necessarily very likely for external hackers. A frustrated or corrupt programmer at the supplier could the controller give "destructive additional functions" as a "gift".

By skillful use of preconditions, however, even these attack vectors can be avoided.

Basic Settings

This functions purpose is for training routines of vehicle-specific values. These typically includes limit stops[62] or sensor calibrations[63].

For this purpose these routines are only startet. After the start there is no further diagnostic communication needed. In the production line unplugging of a off-board diagnostics after starting is quite convenient because it allows the automatic and unobserved execution of the training routines.

These routines may be potentially dangerous if actuators are driven. Secretly started unintentional movements of actuators can be started, that could cause severe injuries.

Even faulty calibrated sensors can lead to negative effects in the vehicle. The impact can range from annoying the driver to the damage of the vehicle.

9.3.3 Flashing

A very complex field for OEMs and hackers alike is flashing ECUs. After all, techniques and know-how of diagnostics, specifics of bus systems and microcontroller-specific methods and tools are relevant.

That does not exclude abusive flashing as an attack scenario.

[62] e.g. window lifter

[63] e.g. steering wheel sensor, yawrate sensor, acceleration sensors

Bootloader

The bootloader are the first lines of code that are executed by powering of an microcontroller or ECU in our case.

This small programme provides all necessary data[64] to jump to the application software or preparation for flashing.

The bootloader is not in any case flashable by diagnostics. Nevertheless, again the frustrated programmer may be mentioned who could have become active in a destructive manner.

Application Software

The application normally contains the operating system, the specific hardware driver's and the controller function models to perform the desired functions.

Targeted changes in the application software are more likely not to be expected. Exceptions are actions of employees at the supplier of ECU, software or function.

In some cases, the application software not only includes the formulae of the regulator, but also their parameters[65].

Changes in the control parameters here are possible in principle, but would require a very detailed background knowledge of controller function, parameter setting and memory management of the controller.

A complicating factor is that the flash process normally is protected by various mechanisms such as encryption, certificates, hashes, etc.

Parameter Set

With many control units, the parameters of the control algorithms are grouped in separately flashable data packets.

---

[64] memory adresses, basic diagnostics etc.

[65] ECUs or intelligent sensors with few, not externally adjustable or fixed parameters.

Advantages of separately flashable parameter sets such as an easier adaptability within vehicle platforms, and simplified validation by testing.

In contrast to the application, the parameter set is partially waived particularly complex encryption mechanisms, which can facilitate reengineering well.

The effort to modify parameter sets[66] and finally flashing of the ECU is relatively high. The damage by overwriting the memory with random values in case of a weak protection is still possible.

9.3.4 Summary - Writing Access

The range of possible attacks by writing accesses to the control unit via diagnostics is very high. The good news is that except attacks by internal troublemakers all write accesses can be secured very well.

The only obstacle here is the knowledge of potential abuse and their assessment regarding likelihood, technical effort and impact.

As in many other issues, a balance between comfort and safety must be found.

[66] purposeful modifications

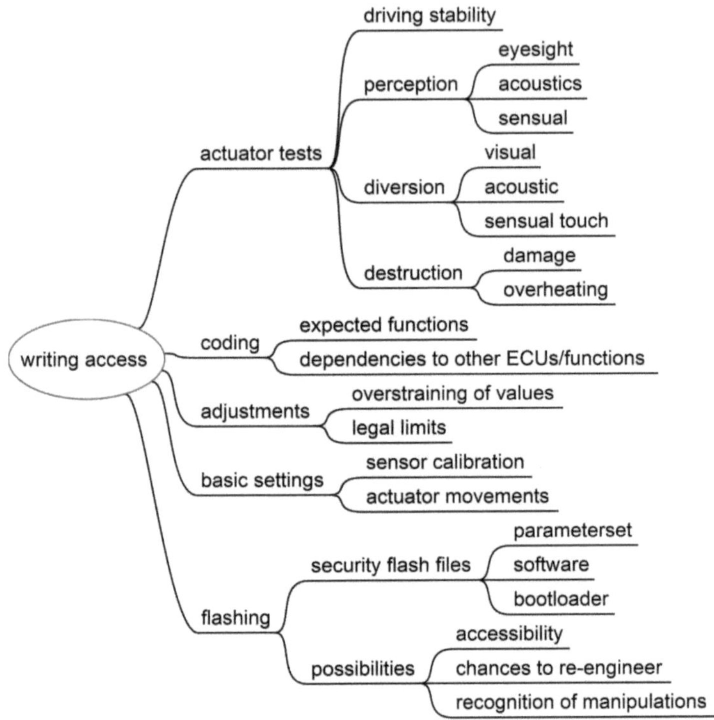

Fig. 63.: critical aspects of writing access

9.4 Summary - Applied Diagnostics

Applied diagnostics offer a variety of potential attack scenarios. It may represent the simplest and most effective way of influencing the vehicle by hacking.

But there is hope.

Following a few basic rules in development of the applied diagnostics gives a chance to set up a decent protection against these attacks.

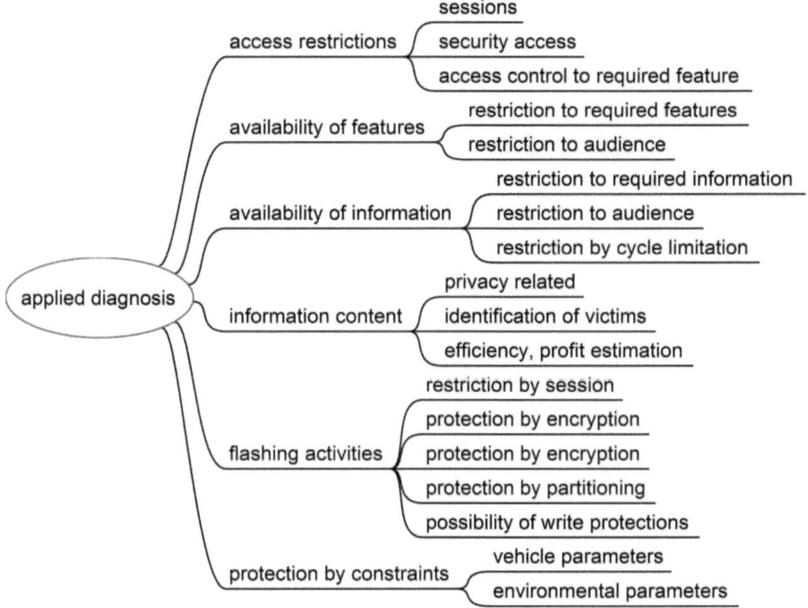

Fig. 64.: protection mechanism for applied diagnostics

The current diagnostic methods are already equipped with a variety of appropriate functions.

9.4.1 Utilisation of full range of functions

Convenience is a bad counselor in terms of safety and security. It is almost negligent not to use the available diagnostic methods in its entirety.

These diagnostic methods do not have to be developed from scratch, they are usually included in the operating system[67]. Therefore it is only necessary to apply the methods through parameterisation[68].

[67] following automotive standards like AUTOSAR [25]

[68] anyhow it requires effort

Avoiding the use of latest diagnostic methods in customer service or production line sooner or later leads to attacks on vehicles, warranty costs and image loss.

9.4.2 Building barriers

There are a number of limitations that may be used. It starts with restrictions for user groups[69] and ends with restrictions by constraints.

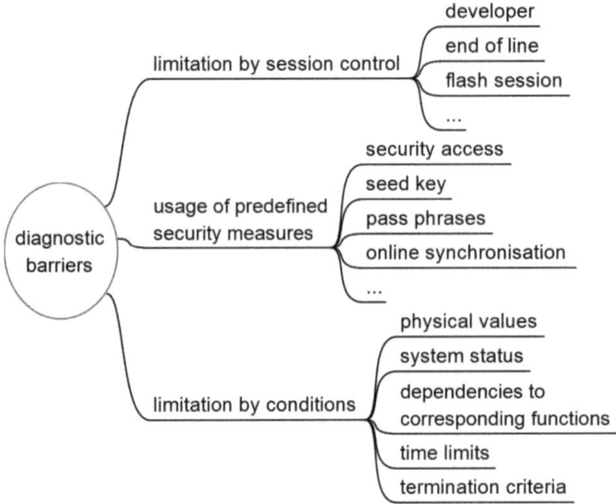

Fig. 65.: samples of diagnostic barriers

User groups

The consideration, which user is allowed to use which functions must be done very carefully. It is not the question, which function **can be used**, but which function **is mandatory**. The keyword here is the usage of **sessions**.

Each session can perform various functions and security mechanisms can be assigned. If in doubt, a restrictive approach is always beneficial.

[69] customer service, production line, developers, software updates

Security methods

The selection of the various security methods has a great influence on the chance to be hacked. Encryptions only make sense if they are implemented with sufficient strength.

Pin numbers provide virtually no protection. They are available online[70] within the shortest time.

Usuage of conditions and constraints

Sometimes it is necessary to implement diagnostic methods with potential risk[71].

In these cases it is advisable to limit the execution of these diagnoses to conditions or constraints. At low speeds for instance, actuator tests would have less impact in case of abuse and would be much easier to control also for inexperienced driver's.

The reduction of actuating variables of these tests[72] can prevent worse.

The shutdown of diagnostics while driving[73] is also possible and already implemented in many cases. Likewise, tests may be limited to certain user groups[74].

Restrictions by operational time of tests is also possible and can help prevent damages to the vehicle or health.

9.4.3 Reduction of Diagnostic Options

One of the most effective ways to defeat hacks, is preventing of the implementation of risky options. Only an option which is not implemented cannot be abused.

[70] forums, usenet, relevant hobbyist websites, hacker and developer communities

[71] actuator tests, basic settings

[72] reduced volume for speaker tests, reduced brighness for head lamp tests, limited motor speed for actuator tests

[73] no or limited diagnostics access at speed > 0 km/h

[74] ESP valve tests only in production line

The KIS principle[75] not only reduces the development effort[76] but also provides time and convergence for developing the absolutely necessary functions with a maximum of accuracy.

9.4.4 Destructive thinking

Thinking like hackers already in phases of pre-development can show weaknesses of the system and the diagnostic options sufficiently early. This can help to prevent this attack routes or at least to impede.

[75] keep it simple

[76] complexity, development time and costs, hardware and performance requirements

10 Network Communication

In modern vehicles, a variety of control devices is installed. Very often functions are no longer allocated on individual ECUs, but on several ECUs. Therefore, cars now rolling are computer centers with a complex network architecture.

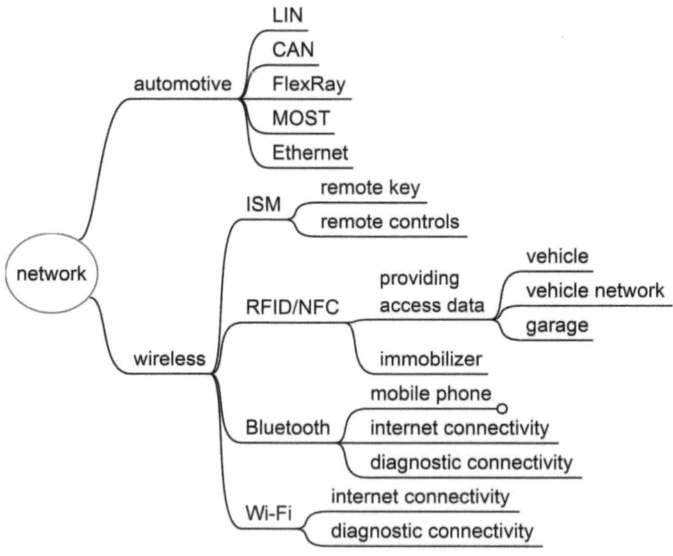

Fig. 66.: vehicle network systems

Recently, the development of the network architecture was mostly guided by technical and safety[1] requirements.

- What speed is needed?
- What quantities of data have to be transported?
- How important is the availability[2] or determinism[3] of the data?
- Is it necessary to send redundant data for safety reasons?

[1] functional safety guided by ISO 26262

- How expensive can it be?

Data security itself has not really been taken into account.

For all networking options, the misuse of each opened data channel should be considered.

10.1 Automotive Bus Systems

Today's bus systems recently used were developed for a function-oriented and lightweight ECU communication.

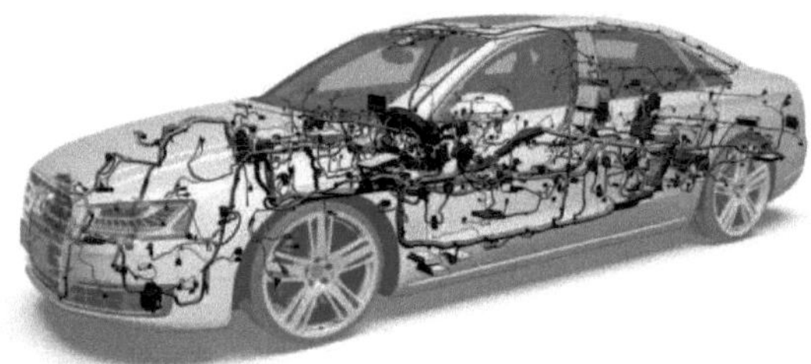

Fig. 67.: Audi wiring harness [20]

There are several different automotive bus types used in vehicles depending on the given aspects.

10.1.1 LIN - Local Interconnect Network

A very simple bus system is the LIN. This cheap and slow one-wire bus is mostly used for connecting external sensors[4], switches[5] or simple actuators[6] to other ECUs. The limited number of cross linkable ECUs is reducing the number of applications.

[2] possibility of loss of data

[3] predictability when data are available

[4] rain-light-sensor

[5] light switch, pitman arms, simple clima control units

[6] air condition flaps, fans

The protocol of LIN is very easy and does not contain mechanisms for functional safety, security and certainly not for data security.

The LIN is only protected against hackers by the lack of popularity[7] in the computer scene. That could change by a rising popularity and prominency of LIN for home automation.

A manipulation of sensor data[8] or activating actuators[9] are just two examples of attacks by using the LIN bus.

10.1.2 CAN - Controller Area network

The CAN is one of the oldest automotive bus systems and is used by most of the applications realised in the car.

The applications vary from crosslinking of powertrain ECUs[10] with comfort systems to network of chassis control systems.

CAN is a very flexible and highly configurable bus system. The CAN protocol makes it possible. Just like the LIN, the CAN is not really designed to protect against hacking.

However, the complex structure of the communication matrix[11] and the PC atypical interface adaptors represent a knowledge barrier for potential attackers.

In case of successful reverse engineering and the spreading of this information, the options for attacks are wide ranging. Manipulations of neccessary sensor data, actuator requests etc. are also feasible.

The system-related timing tolerance of the CAN bus and the relatively low data rate unfortunately allow the use of gateways. These gate-

[7] so far untypical interface adaptor

[8] feign light situation to supress switching on headlight, supress rain information for switching off wipers

[9] closing clima flaps or stopping fans (fogging, overheating), driving seat adjustments

[10] engine ECU(s), gear ECU

[11] list of the transceiver relationships; which ECU is sending which signal and which other ECUs are using them

ways do not set high standards for the microcontroller and programmer. Cheap 8 bit processors and an open source IDE[12,13] are sufficient, if the signals to be manipulated are identified.

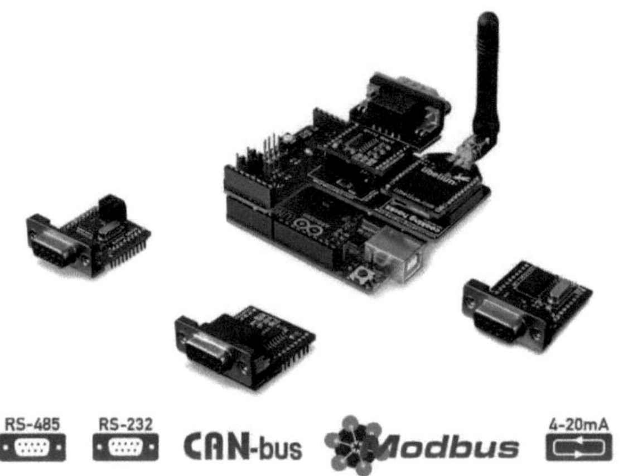

Fig. 68.: toolset for several bus systems based on Arduino [31]

Like the LIN the CAN enjoys certain popularity in the field of home automation, which increases the availability of potential interfaces.

10.1.3 FlexRay

Especially in the field of driver assistance systems FlexRay has already replaced the CAN. In premium cars, FlexRay is most commonly used for chassis control systems and Powertrain. The low-budget cars will follow that trend within the next years due to falling prices of FlexRay components.

FlexRay has a lot of advantages compared to CAN.

[12]integrated development environment
[13]e.g. Arduino

A high throughput, a large amount of data and the predictablility of availability of information make FlexRay role model for driver assistence systems.

Although the FlexRay protocol is constructed quite simply, the large amount of various signals and timings make direct attacks significantly more difficult.

Reengineering is further complicated by the high degree of specialisation of the bus system and thus a low availability of low-cost interface adaptors.

The installation of self-developed gateways for signal manipulation is not excluded in principle, but it requires powerful processors[14] and high detailed knowledge of the configuration of the interface driver ahead.

10.1.4 MOST - Media Oriented System Transport

An even higher specialisation on infotainment applications contains the MOST bus.

A very high data rate, rather exclusive transceiver modules and a complex protocol with a relatively small intervention option make the hack via MOST unattractive.

The illegal copying of copyrighted material[15] by use of the MOST is conceivable but it requires an extraordinary effort. A direct ripping of the disk on personal computer is much more simple and efficient.

For adding further media sources[16] MOST units are available in aftermarked. These components work as gateways between those audio and/or video sources to MOST.

---

[14] 32 bit controller

[15] DVD or Blu-ray content

[16] DVB-T (TV), additional media player (HiRes audio player), Blu-ray player, gaming consoles

The customer may not necessarily comprehend, what other functions[17] are implemented by the manufacturers[18].

Even if an attack on the bus system itself is unlikely, due to their high crosslinking potentail, infotainment control units are interesting targets.

10.1.5 Ethernet

A relatively new development in automotive technology is the use of Ethernet. The aim is mainly to use the high speed and the low component prices for use in the vehicle. This, however, opened a door in the computer world that literally invites hackers, because experience in this area from PC, Server and mobile devices is detailed.

The Ethernet protocol, the interfaces[19] are the staff of life in the Internet scene.

The replacement of MOST[20] or FlexRay[21] by Ethernet might be a commercial advantage, regarding security, however it is a disaster as well as leading out of Ethernet from the closed and protected environment of the vehicle network to diagnostic interfaces.

What serious intervention an nearly unprotected access to diagnostics for hacking attacks provides is described in chapter 9 - Applied Diagnostics on page 84.

10.2 Data Routing

The bus systems in the vehicle already described are not separated data channels. Information available on a specific bus system, may be needed and used in other bus systems.

[17] perhaps undesirable

[18] partly no known brands (no-name)

[19] personal computers are equipped with Ethernet ports, evaluation boards like Raspberry Pi and clones, too

[20] infotainment applications

[21] driver assistence systems

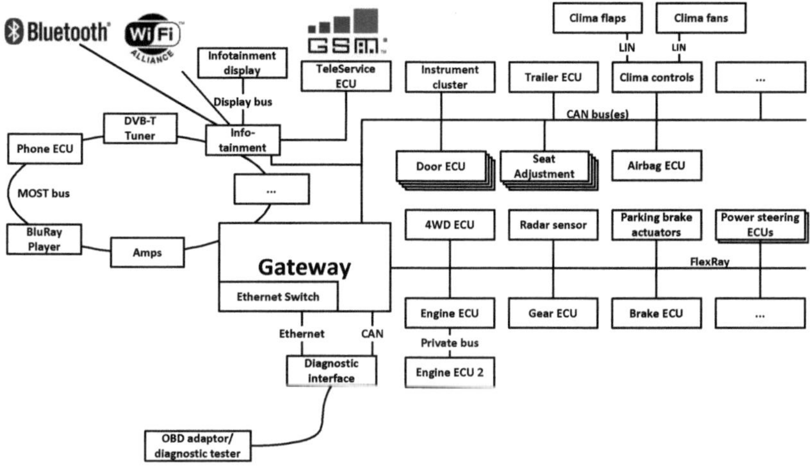

Fig. 69.: typical topology of a car

In the Figure 69 a highly simplified topology of a car is shown. Usually, significantly more bus systems and control units are fitted in a real vehicle.

In principle information is only buswide available. If an ECU needs data that are provided on other buses, they must be routed from a gateway. The control unit that performs this task is the gateway.

Therefore the gateway is an implemented "man in the middle", which is awaiting hacking manipulations.

10.2.1 Diagnostic Communication

Not only signals have to be routed, but among other things, the complete diagnostic communication that takes place on a separate bus.

So what is more natural than to manipulate the routed diagnostic handshake communication between ECU, gateway and diagnostic tester unit?

The intervention options offered by diagnostics will be considered in chapter 9 - Applied Diagnostics starting on page 84.

10.2.2 Signal Manipulation

Manipulating data[22] is normally prohibited by development policy at the OEMs. However, this should not prevent hackers to make changes in the data in case of access to the gateway.

Particularly good protection against manipulation[23,24] of the gateway is essential.

10.3 Wireless Bus Systems

But there are not only wired connections used in cars. The number of applications for wireless communication is growing continuously. And these wireless networks partly have access to the vehicle network!

10.3.1 Industrial, Scientific and Medical Band

Since the introduction of radio keys, the Industrial, Scientific and Medical Band (ISM) is used in the vehicle. It draws on specifications of Short Range Devices (SRD) and uses typical frequencies of 433 MHz or 866 MHz[25].

Typical automotive applications for ISM are

- radio keys
- garage door opener[26]
- remote control for auxiliary heatings[26]
- tire pressure monitoring system[26]

A long-term use does not protect against misuse, especially if it is possible with simple and cheap technical means, at least to interfere in the communication. Overlaying a closing command of a remote control key by easily modified PMR[27] handheld radio is a simple exercise.

[22] signals

[23] flashing new software, change of parameters

[24] see subsection 9.3.3 - Flashing on page 101

[25] country-specific

[26] optional equipment

[27] Private Mobile Radio

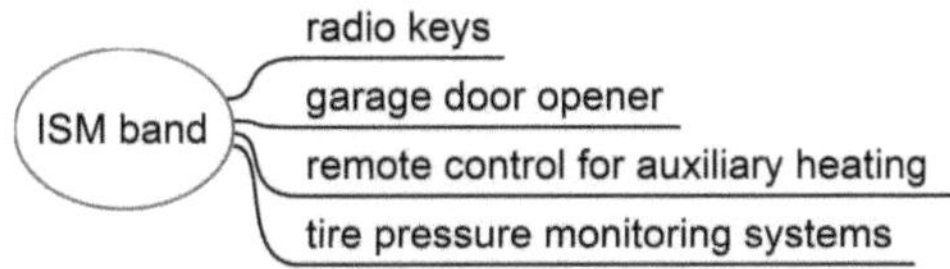

Fig. 70.: typical applications ISM band

Sometimes disturbances are quite sufficient to override convenient functions.

An expected attack on vehicles would be burglary up to the theft of the vehicle.

10.3.2 Radio-frequency identification, Near Field Communication

One of the first applications for radio-frequency identification (RFID)[28] is the immobilizer, where a very short-range radio system is exchanging crypted access codes between a built-in chip in the car key and the vehicle.

The very short-range working principle seems to be safe against attacks, but it is not.

Just recently the RFID technology loses its quite good protection state which has been given for years by its wide application and thus disseminate information.

Significantly newer than RFID is NFC, which e.g. can be used for the transfer of configuration data[29] to smart phones, badges and other personally authorised devices.

Programmable NFC modules are easy to obtain. The configuration is done via smart phone. Modified login data to vehicle network or Internet links to viruses and Trojans could be installed in smart phones very

[28] also based on the ISM band

[29] Bluetooth or Wi-Fi access data (PIN, SSID), car mode for smart phones

easily and without arousing suspicion. The possible range of options is known to true hackers very well.

10.3.3 Bluetooth

Already in the vehicle for a relatively long time is Bluetooth in the typical application of hands-free kits for mobile phones. Over time, however, the profiles[30] used have changed.

Fig. 71.: typical Bluetooth applications

Vehicle Integrated Units

In the beginning of Bluetooth in the car the Hands Free Profile[31] (HFP) has been used. This quite simple protocoll only streams audio data for telephone application. To hack this stream does not really make a lot of sense.

The hacking of SIM cards as part of the cellular phone with phonebook, adresses and access data to mobile network is much more interesting. The usage of the driver convenient SIM access profiles opens up new possibilities[32].

[30] application specific protocol

[31] profile = application specific protocol

[32] (theoretical) options of open data channels

The development of the Bluetooth protocol and the associated profile has also led to new applications in automobiles, especially in infotainment systems. So audio streaming[33] and Internet connections[34] via Bluetooth have been established.

While audio streaming provides little attack scenarios, the entire Internet access represents a whole range of possible attacks, in particular because infotainment systems are often based on embedded operating systems[35] like Linux[36], embedded Windows[37] or proprietary like QNX® CAR Platform[38].

The development of applications (apps) for these operating systems has great popularity and thus knowledge of the techniques and methods of these operating systems has achieved a widespread use.

The confusing large market not only offers obvious automotive-related apps. For bad apples the abilities are also given to accommodate various automotive functions in seemingly unproblematic apps.

A quite new application is the use of iBeacon[39] units for setting up smart phones. This standard deliberately avoids a transceiver authentication. An idea is a convenient transmission of configuration data to smart phones or smart devices like access data to Bluetooth or Wi-Fi access points of the vehicle.

What speaks against the opening of additional data channels by modifying or replacing iBeacons, which are available on market for little money?

---

[33] Advanced Audio Distribution Profile (A2DP)

[34] Personal Area Networking Profile (PAN)

[35] embedded distributions for Systems on Chip (SoC)

[36] e.g. GENIVI, TIZEN In-Vehicle Infotainment (IVI)

[37] Windows CE, Windows Embedded Compact, Windows Embedded OS

[38] a Subsidiary of BlackBerry

[39] standard for indoor navigation

Via Bluetooth wearables also could be connected to the vehicle for driver monitoring. A misuse by insurers and secret services cannot be excluded here.

Bluetooth Adaptors

A relatively new trend is the use of cheap OBD adaptors based on Bluetooth or Wi-Fi, often made in China. However, these adaptors provide direct access to the vehicle network.

In combination with respective apps several diagnostic commands can be sent via adaptor into the vehicle network. These commands could be used for attacks. Further details about attacks via diagnostics see chapter 9 - Applied Diagnostics on page 84.

Hardware and software from obscure sources not only entails risks by use of Apps, but also by possibly faulty or deliberately manipulated devices.

Cheap OBD adaptor set substantially to also inexpensive components[40] that are described in great detail in various forums and accordingly easily programmable.

The opportunities for hackers to be destructively active, are terrific.

10.3.4 Wi-Fi

A significantly better performance than Bluetooth offers Wireless Fidelity (Wi-Fi), which is also known from the personal computer.

Wi-Fi offers higher transmission rates and also other interesting features [41]. Some interesting features for automobiles are considered below.

Especially multi-purpose interfaces such as Wi-Fi open many opportunities for vehicle manufacturers, which may be of course well advertised by marketing. But here there cannot be enough warnings not to un-

[40] microcontrollers partly Arduino based, standard Bluetooth modules with programmable controllers

derestimate the risks in the use of options, for example crowdsourced hotspots.

Wi-Fi Aware

A small excerpt from the website [42] features shows possible applications, some risks in footnotes.

- Provides continuous device-to-device discovery, even without a GPS, cellular or hotspot connection[41]
- The technology works well in crowded environments and indoors - by design
- Wi-Fi CERTIFIED means it operates in typical Wi-Fi range across numerous device brands
- Allows bidirectional sharing of small pieces of information, e.g. localisation data[42,43], sensor readings[44], and services in proximity
- Applications easily switch[45] to a Wi-Fi connection to use high bandwidth
- Wi-Fi Aware is app-centric, and users can control privacy[46] and opt-in or out of identity disclosure
- Leverages the spontaneous nature of user proximity to enable innovative applications

The options of this feature should be considered critically and analysed for the benefit and potential risks.

Wi-Fi Passpoint

With Passpoint a hotspot functionality is provided. Unchecked turning on data channel to the Internet without a separation[47] to vehicle network can enable massive attacks.

[41] useful in case of hacking for permanent observation of the target person
[42] NMEA data
[43] even in GPS free car parks
[44] addition of vital signs to vehicle data
[45] possible lack of security
[46] only if the user ever sees what is meant, technical overtaxing the user
[47] firewall

Wi-Fi WMM Programs

WMM stands for Wireless Multimedia Extensions and contains low tolerance for latency, packet loss and jitter for using voice, audio or video applications. It also contains hotspot functions.

Attacks using compromised data seems to be sufficiant for disturbing the vehicle network in case of to week firewalls.

Wi-Fi Direct

Wi-Fi CERTIFIED Wi-Fi Direct® offers direct connections between devices in a simple and convenient way.

Simplifications and convenience often contain the risk of ignoring the security and safety concerns. The implemented protection by establishing a data connection for devices in any direction is implemented is not clear at the moment.

Diagnostic Interface

As already shown, WiFi provides fast transfer rates of partial simple establishment of data links. The Wi-Fi interface to the vehicle is predestined to accept the diagnostics, is already used in some OEMs Ethernet for diagnostic buses.

It sounds appealing to be able to read out the fault memory and the mileage already when entering into the garage. However, it must be clear that diagnostics is always a two-way-communication and this fact must not be forgotten or underestimated.

Car-to-X Communication

Extremely interesting for hackers could be the car-to-x[48] communication.

[48] including car-to-car communication

Not only the car-to-car communication itself offers targets for attacks. Freely accessible residential facilities used for car-to-x[49] could be abused for attacks quite easily.

Emergency brake warnings from its own vehicle or green signals from a manipulated red traffic light[50], could be enough to cause at least rear-end or side impact.

Further attention has to be paid, that not all vehicles will be equipped with car-to-x in the future and thus human erratic behaviour must be reflected in the consideration.

10.3.5 Mobile Radio

A wonderful unleashing of local data networks is the implementation of mobile communications. Of course there are a lot of potential applications. That these applications, however, also involve risks, is often neglected.

As mentioned repeatedly, any data link can be misused in the Internet or telephone network, and represents a non-negligible gateway for potential hackers.

Telephony

The phone in the car is of course a great thing. Nevertheless, more and more countries regard it as a great distraction[51] from driving.

The course also offers telephony risks outside distraction. A parallel transmission of data packets without speech content is just as conceivable as manipulations of volume or the import of noise[52].

In some criminal imagination various scenarios are conceivable. In addition to noise or fright by volume increase and the forwarding of the audio stream to third parties is possible.

[49] e.g. Wi-Fi modules mounted on traffic signs or traffic lights

[50] shielding traffic light antenna, replacing by own Wi-Fi sender

[51] in spite of the use of hands-free kits

[52] see subsection 4.2.2 - Acoustic Effects on page 50

Also the abuse of installed phones in the vehicle and the mobile phone microphones represents a potential risk. The interception of communications could be realised in this way and the (mis)used communication paths would be obscured pretty well.

Mobile Data Services

Mobile data connections can expand the radius of the attack capabilities of Bluetooth and Wi-Fi in an almost unlimited way. The only limit is the network coverage of the phone service provider.

Teleservices

A fairly new field for the carmaker is introducing TeleServices.

Fields of application are, for example, suburban aids in case of damage[53], reading the failure memory[54], hardware and software releases for product recalls or silent software updates[55].

Other possible applications may be more specific service messages or messages analogous to traffic messages channel (TMC) services.

While unmotivated service calls are annoying at best, and possibly give rise to costs, manipulated traffic messages can lead to targeted route changes. The intentions are not really clear.

Theft Protection

Satellite based theft warning systems are quite common in some countries. Who says, however, that the combination of high-precision GPS data and open data channels are not misused?

The tracking of the target population or creation of movement profiles by insurance[56] are hereby possible.

[53] increase the availability in case of failure

[54] failure statistics

[55] avoiding recalls of software errors, automatic corrections or functionality adjustments

[56] criminal motivations of insurance are naturally excluded

10.4 Summary - Network

The communication among the ECUs with each other as well as the additional crosslinking of the vehicle to the Internet is opening a plethora of opportunities for hackers to damage the ECU, the vehicle and, at worst, inflicting injuries to the driver and passengers.

10.4.1 Current Situation of Series-Production Vehicles

Manipulation of the in-vehicle bus system are largely only on possible with the help of changes in the ECU software or the installation of additional hardware[57].

Currently, the security of communication is largely limited to bus-specific control mechanisms, which are usually supplemented only for safety-critical signals with context-bound security measures.

10.4.2 Situation in Future Vehicle Models

Previously this was not a big lack. The distribution of functions that require multiple ECUs such as autonomous driving requires a new level of communication security. ECUs with actuators must accept commands from other devices without the possibility of testing for plausibility and fulfil these commands. This communication capability has been opened in recent years of development and will be used in the next model series effectively.

Security for Actuators

An **highly encrypted end-to-end authentification** in addition to the bus specific security measures at least **for commanding of actuators** is highly mandatory.

[57] selfmade gateways

Security for Sensors

Sensor information on buses should be examined regarding **to privacy** and misuse by third parties using **appropriate encryption measures**[58] prevented.

Security for Diagnostics

Looking at the cross-linking as a basis for the diagnostics is thinking about other safety-critical aspects.

As for purely practical reasons, there will always be a certain number of required diagnostic functions. It is advisable to distinguish between tending uncritical diagnostic functions[59] and functions with high hacking potential[60].

For **critical functions encrypted authorisation methods** should be considered.

10.4.3 Expansion from vehicle network to Internet

As already mentioned providing data channels to the Internet – consciously[61] or unconsciously[62] – is an enormous risk for security and safety.

The current network topologies of vehicles should be mandatory investigated for potential interventions and future topologies protected accordingly.

This could mean for the future a considerably intensified authentification between the signal transmitter and receiver as well as an extension of the vehicle gateways to **firewall functions**.

---

[58] non static end-to-end encryption

[59] e.g. read out identification data

[60] e.g. driving actuator tests, reading out sensitive position data

[61] telematic or telemetric systems, car-to-x communication

[62] Bluetooth or Wi-Fi connections to mobile devices

10.4.4 Wireless Networks

Wireless networks are very convenient. However, this convenience is a clear risk for safety and security, especially when interactions do not necessarily require a lot of background knowledge and technical know-how.

To make wireless applications rudimentarily safe the **detection and treatment of interferers**[63] must be given as well as **protocol-based attacks**.

The creation of additional networks via further wireless hardware[64] allows the existence and abuse of further security holes.

[63]PMR handheld radio, jammer for Wi-Fi and Bluetooth
[64]e.g. smart watches, wearables

Part IV.

Defensive Measures

Intuition in predicting hacking attacks on vehicles needs a lot of background knowledge in various aspects, aspects which have been illustrated in the previous chapters.

Now necessary questions have to be asked in a structured way to avoid hacking.

11 Technical Measures

The first thoughts on vulnerability of a car should be made in the phase of planning[1] of features. Detailed analyses and according technical measures should actually be mandatory to remove weaknesses.

The following activities are not completely covered by existing development processes and have to be supported by organisational substructures[2].

11.1 Vehicle Level

For many engineers in the automotive world looking at the most general vehicle level is almost an affront. They prefer developing tricky solutions of technical details and tend to over-engineer. However, the view on vehicle level is extremely important.

Only a very general view on the vehicle's features can demonstrate the extent of the planned features as well as the undesirable side effects.

11.1.1 Feature Description

For this purpose it is necessary to describe the desired effects and their dependencies at the vehicle level in a very abstract way – as required following various product development plans.

A good approach is, to go through each step from arriving at the car, entering and use of the planned feature mentally and document the process.

Each step should be checked for relevance in respect to the planned feature. In case of missing relevance for the feature the answer and its reason should also be kept in the records.

---

[1] concept phase

[2] see section 12.5 - Organisational Effort on page 153

A standardised situation catalogue could be helpful to avoid forgetting any circumstances. For sure, hackers will recognise and exploit many more situations. Thinking destructively should not be left to criminals.

The sample shows in **??** the putative triviality of this examination level. Similarities with the classification tree method are not random.

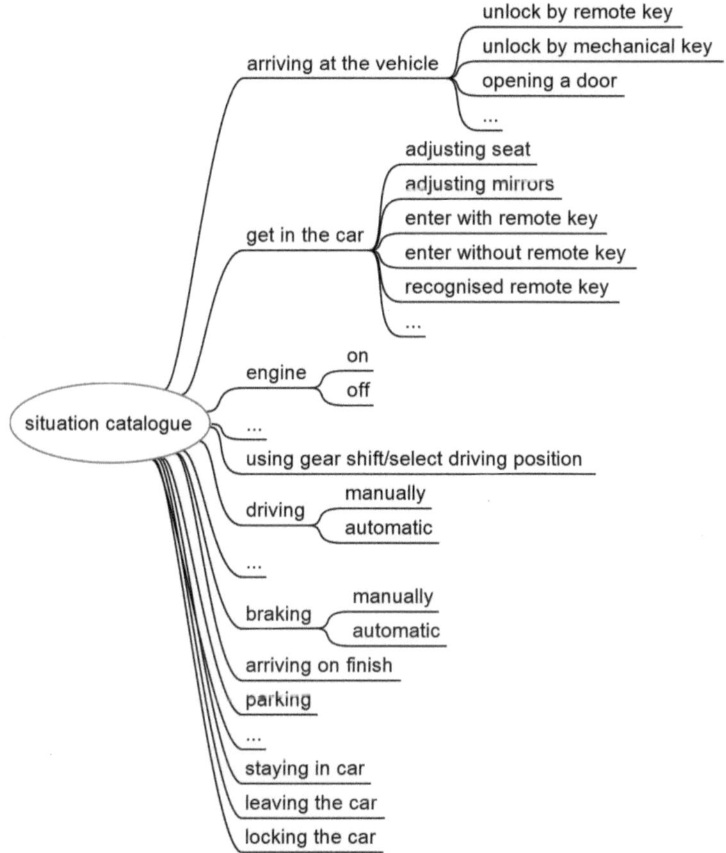

Fig. 72.: rudimentary situation catalog

Despite the simplicity this task should be executed very carefully.

Mission 1 *Detailed description of the feature with all cases of use and effects from a user perspective on vehicle level.*

For the analysis of the cases of use the complete product life cycle from first idea to scrapping[3] should be considered.

The detailed description of the use cases should identify all concerned persons.

Mission 2 *Preferably complete identification of users, also the implausible persons in the beginning.*

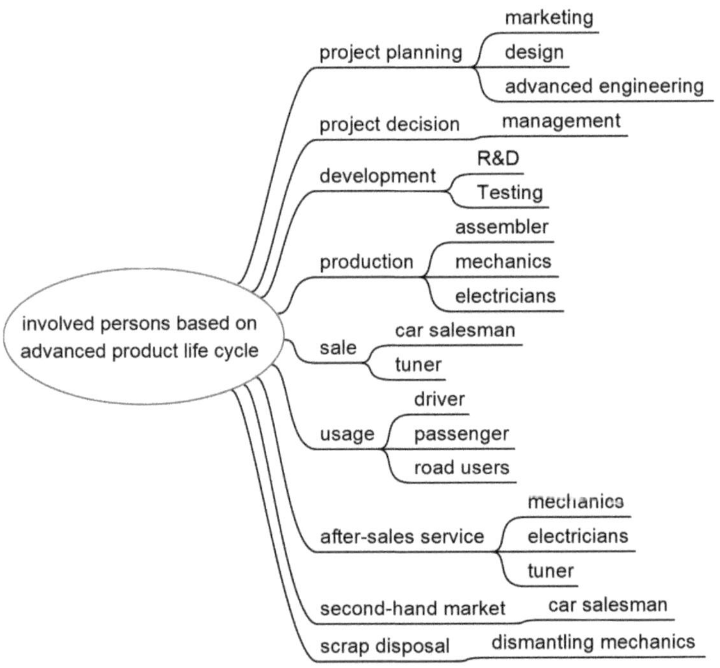

Fig. 73.: concerned persons based on advanced product life cycle

[3]see Figure 73 - involved persons in product live cycle on page 132

An identification of the involved persons will allow a deeper look on the motivations[4] of the normal user. Out of these considerations the rated situation catalogue should be reviewed and reworked.

11.1.2 Feature Effects Analysis

Features usually consist of a sum of different effects that manifest themselves in different ways to the user. They addresses several senses or uses different and work following indipendent physical principles.

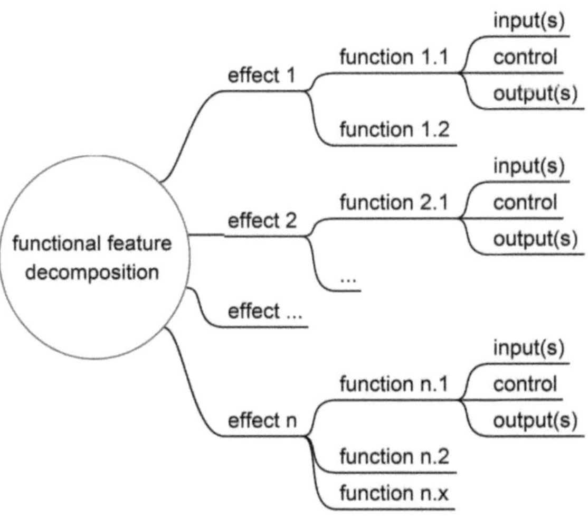

Fig. 74.: feature decomposition

A decomposition of the feature into these different effects have to be done.

Mission 3 *Decomposition of the feature into effects using indipendent physical principles.*

The next big step is the analysis of the desired effects on possibilities of unforseen use or misuse.

[4]see chapter 3 - Motivation of Hacking on page 30

Results from these analyses may well lead to further iterations by themselves.

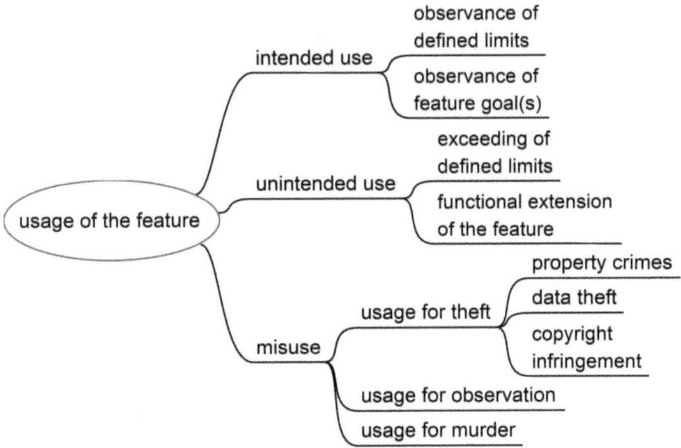

Fig. 75.: usage possibilities of the feature

Mission 4 *Detailed analysis of usage possibilities of the decomposed effects including misuse.*

This analysis requires a lot of knowledge and experience, also in the respective subregions the product life. Therefore perhaps, this analysis requires organisational measures[5].

11.1.3 Decomposition of Effects

Also for the next step a lot of technical and functional background knowledge is required.

The perspective now is changing from abstract effects of the customer feature to its required functions and subfunctions[6].

At first the feature should be decomposed based on the effects.

[5]e.g. department with cross-divisional function

[6]see Figure 49 - feature vs. function (II) on page 72

Mission 5 *Decomposition of the effects into its functions on vehicle level and its inputs, controls and outputs.*

This first consideration should include only the for the feature compulsory required functions. The use of the theoretical control loop[7] can provide a good basis for this.

Translated to the functionality of a feature a first decomposition could be to put outside the inputs[8], the control[9] itself and the outputs[10]. Now the image of the functional decomposition in Figure 74 is complete.

11.1.4 Vehicle Network Architecture

In the recent time network architecture was mainly reviewed or adopted in case of increased network communication[11] or changes in technology[12]. This will not be sufficient in future any more.

Of course, the introduction of new network buses is a good reason for an intensive review. New technology provides a lot of new methods and possibilities.

Analysis of the network architecture should be no single event in the development of a vehicle platform. It should be included into a permanent iteration loop, considering technical evolutions and hacks or uninvited modifications, that appear by time.

Availability of Information

In the past many manufacturers trust in a lack of information for hackers. The installed bus systems and their configurations have been a "secret" for several years. But the numbers of involved people has been

[7] see Figure 53 - closed loop control on page 77

[8] see section 8.1 - Sensoring on page 77

[9] see section 8.2 - Control Algorithm on page 79

[10] see section 8.4 - Actuators on page 79

[11] performance gap

[12] replacement of CAN by FlexRay or MOST by Ethernet in some domains

increasing permanently. The more suppliers[13] the lower the chance of keeping the secret of network communication.

Manuals of components of the aftermarket[14] partly expose further information about the vehicle network topology. In the end, necessary information is available in the Internet quite fast.

Mission 6 *Never trust in security of information.*
Confidentiality requires sufficient encryption.

Fig. 76.: DIY CAN monitor [12] based on Raspberry Pi

[13] Tier-1, including second source suppliers, sub-suppliers, consulting engineers, test labs etc.

[14] e.g. trailer control unit [47]

Complexity of the Network Architecture

A big challenge is the development of the network architecture in the vehicle regarding functional and safety requirements and now the new topic of security has to be added, too.

The increase of complexity can be a good defensive measure, but it can open a lot of new possibilities for attacks.

Mission 7 *Reducing complexity to manageable levels.*

Low complexity in combination with older bus systems in most cases is quite weak against attacks [28] [50].

A high complexity combining older and newer bus systems is very hard to analyse. This is where two extremes finally meet: a lack of implemented security measures[15] in the older bus systems and a lack of experience with the newer buses.

Mission 8 *Complementing existing network protocols with additional protective measures; detailed analysing new bus systems with respect to possible attacks.*

11.2 System Level

After analysing the vehicle the system whose aim is to perform the feature and its functions has to be considered. All necessary components[16] and their relationships[17] are in focus now.

11.2.1 Location and Allocation of Inputs and Outputs

In the previous steps the feature has been decomposed into its effects, functions and functional subsystems.

[15] could be partly compensated by skillful complementary measures

[16] as a black box

[17] architecture, communication

Features and their functions are based increasingly on existing components. In many cases they are purely developed as software modules. The task now is to locate existing inputs and outputs and possibly reassign to ECUs.

Mission 9 *Identification of existing inputs and outputs, allocation of missing functions.*

It should not be forgotten that various aspects have to be considered, such as

- sufficient accuracies[18] and effective range[19]
- sufficient functional safety classifications[20]
- sufficient security against cross-effects[21]
- avoiding unforeseen additional functions

Mission 10 *Clarification of freedom of interferences to existing features and functions.*

11.2.2 Architecture Analysis

Allocating functions to existing ECUs is not enough in ensuring safety and security of the feature. Very important for the stability and resistance to hacking is the network architecture[22] or topology of the vehicle.

Questions on data rates, cycle times, data volumes etc. are asked in the normal case in the standard development process. The functional safety[23] has top priority, too.

An analysis of the topology of the vehicle therefore must consider a number of issues.

[18] sensors/actuators resolution, value ranges, repetition cycles, signal confidence

[19] actuator strength, working range, cycleability

[20] ISO 26262 and comparable standards

[21] avoiding interferences

[22] see Figure 69 - typical topology of a car on page 115

[23] regarding ISO 26262, including end-to-end protection

Content of Information

The check of the content of the information (data) is a first step to ensure safety and security in case of hacking. Following some points of view:

- necessity of data[24]
- privacy violating content of data[25]
- possibility of manipulation of data
- impact in the case of manipulation
- functional cross-influences to other information

Mission 11 *Analysis of data content on necessity, privacy and impact in the case of manipulation.*

The reduction of provided and used data to a technically necessary minimum should be a top priority.

Mission 12 *Taking appropriate protective measures (e.g. encryption, limiting values, authorisation) for absolutely crucial but vulnerable data.*

The measures used to protect the data must be up to date, because hackers are very knowledgeable. Simple single mechanisms are not really a barrier for attacks.

Information Flow

In addition to the content, the ways information takes, are interesting for avoiding hacks. The availablity of data on different bus systems[26] should be analysed and limited if possible.

MIssion 13 *Limitation of availability of data to technically dedicated networks and installation of appropriate firewalls.*

Although an open architecture expands the technical possibilities, but also offers at the same time vulnerabilities for attacks.

[24]waiver of non-obligatory data

[25]see Table 9.2 - Misuse measurement values for crime on page 94

[26]see Figure 66 - vehicle network systems on page 109

In the future, the technical complexity of gateways by the use of security features[27] is likely to increase exponentially.

Complex ECUs that access multiple bus systems can represent unintended gateway. Especially in these control units a closer look into the internal communication paths is strongly recommended.

Particular attention should be directed on private buses. This communication is often supplier specific and cannot be checked by the car manufacturer.

Mission 14 *Preventing unintended gateways, preferring official data channels over defined gateways and avoiding nor observable privat buses.*

11.2.3 Technology Leaps

———-> Artificial Intelligence

11.3 Unit Level

Analyses of the control unit regarding functional safety are quite common, even now. Organisational processes for this are largely in use.

Mission 15 *Expansion of scope of **intended use** to **use cases in lifetime**.*

Functional safety is currently mostly applied only to the intended use. The scope of the "intended use" must be extended to production of the vehicle[28], customer service, expectable tuning measures and hacking attacks.

Of course, the fight against hacking is like the race between the hare and the tortoise. It is therefore necessary to prevent attacks if possible or to detect unwanted modifications.

[27] firewalls, encryption, authorisation
[28] end of line at OEM

11.3.1 Recognition of Hardware Modifications

To prevent modifications to the harness[29], is an impossibility. Excluding tuning measures by changing electrical values of sensors or actuators needs a high technical effort[30] with associated costs.

Depending on the resulting risk it makes more sense to recognise the changes. Due to these changes adequate responses can be considered and countermeasures be taken.

Mission 16 *Implementation of techniques for recognition and adequate response of hardware modifications.*

Looking for manipulation only in the wiring harness would be as short-sighted as fatal. Of course, interventions and modifications in the ECU require a lot of detailed knowledge in electronics.

However, the copying of standard circuits[31] or circuits of consumer electronics[32] can significantly facilitate reverse engineering.

Mission 17 *Avoiding sample implementation of integrated circuits and copying circuits from consumer electronics.*

Recognition

Technical solutions for recognising manipulations are depending on the physical and electrical working principle of the sensor or actuator. But in any case it requires measurements, which may have an influence on the sensor[33] or actuator[34].

The methods of recognition of manipulations are more or less the same like electrical failures, such as

[29] in case of access to the vehicle

[30] intelligent sensors or actuators instead of discrete ones, complex encryption

[31] sample implementation of integrated circuits in manuals

[32] e.g. circuit design in infotainment systems

[33] e.g. limiting accuracy or measuring range

[34] e.g. sluggish responses of the actuator

- monitoring of voltages and currents
- monitoring of measuring ranges
- gradient monitoring
- offset monitoring

Detecting manipulations of the ECU hardware itself is likely to be very difficult. But thinking outside the box could open new possibilities[35].

In all diligence a sense of proportion for expenses, benefits and risks should be considered.

Responses

Finding adequate responses on manipulations is quite difficult, but they include a classic conflict of goals: *failure condition*[36] versus *availablity of a feature*.

Mission 18 *Clarifying conflicting goals of* ***error handling*** *and* ***feature availability****.*

So, the decision what changes or manipulations are critical[37] is not only a technical one. Answers to the question what changes are considered as abuse are to be given not only technically. A certain openness for tuner can positive marketing effect. So the decision is quite political. Of course top priority is safety.

Feasible treatments on manipulations could be:

- deactivation of the feature[38] or necessary functions
- readjusting of manipulated values
- use of substitute values[39]
- use of physical models instead of measurements[39]
- ignoring of manipulation, use manipulated values

Again, a decision is not only technically founded.

[35]detection of opening of ECU housings
[36]recognised manipulation in this case
[37]limiting safety, security or durability
[38]worst case, annoyance of the customer for availability of the feature
[39]limitations of feature performance possible

11.3.2 Protection Against Signal Modifications

Signal manipulations are quite similar to interventions in the hardware, but the recognition can be implemented by software in most cases. The range of methods is very wide, but a detailed knowledge of the internal and external bus in use is required.

Typical methods are inter alia:

- cyclic redundancy checks (CRC)
- checksums
- measuring range monitoring
- monitoring modules, model monitoring
- sign, gradient, offset monitoring
- message or alive counter
- time monitoring
- plausibility checks

Corresponding to manipulation of signals, the same considerations as hardware manipulations apply[40].

Mission 19 *Encryption of relevant signals with a hazard appropriated strength.*

The vehicle gateway is also responsible for signal handling. Attacks on the gateway are therefore particularly critical and must be prevented.

For this reason the development of the gateway should be given special attention[41] to, which is already given in most cases.

Mission 20 *Special attention in development of gateway ECUs regarding safety, security and diagnostics.*

[40] see 11.3.1 - Responses on page 142

[41] encryption, safety and security measures

11.3.3 Protection Against Memory Modifications

As discussed in section 9.3 Writing Access[42], the range of possible memory modifications is wide.

Because of this wealth of opportunities an effective protection against attacks on the memory[43] of ECUs is not that easy.

Preventing Manipulations

Prevention is better than cure. Avoiding tampering in generally, is probably impossible.

If the paths of the memory manipulations are recognised, measures can be taken. Therefore it is not necessary to reinvent the wheel.

Mission 21 *Identification of attack routes and methods.*

Operating systems[44] often provide a bunch of tools and methods for ensuring data security, but they have to be configured and used in a skilled way!

Mission 22 *Identification of security measures of the used operating system. Skilled usage of resources of the operating system.*

Recognition of Tampering

Another challenge is to detect tampering. For this purpose, the knowledge of the various implemented writing accesses is imperative.

Mission 23 *Identification of preferably **all** writing options.*

Besides the obvious methods such as diagnostics[45], also learning algorithms lead to changes in the ECU memory.

---

[42] see page 97

[43] RAM, ROM, Flash

[44] see subsection 7.2.3 - Operating System on page 75

[45] adjustments, flashing, basic settings etc.

Various methods are available for testing the correctness of stored data. The level of complexity of these methods is ranging from minimal simple when using checksums, comparisation with sample data[46] to computationally intensive mechanisms such as MD5[47] or MD6[48].

However, it must not be forgotten that ECUs typically are embedded systems with limited computing performance and storage capacity[49]. Therefore, once again, a sense of proportion[50] is required in the use of the methods.

Mission 24 *Choosing **appropriate** and **feasible** methods for testing of the integrity of stored data.*

11.3.4 Protection Against Diagnostic Activities

The protection against attacks diagnostics should hardly pose problems, many measures are already included in the diagnostic protocols and methods[51].

The use of these methods often depends on experience[52] and the scope[53] of the developer.

Mission 25 *Skilled usage of the implemented measures of the diagnostic protocol.*

[46] online synchronisation via customer service or telematics
[47] message-digest algorithm; cryptographic hash function 128 bit
[48] cryptographic hash function 256 bit
[49] cost reasons
[50] assessment of risks
[51] see section 9.4 - Summary - Applied Diagnostics on page 104
[52] diagnostics are often cross-divisional functions
[53] function developer vs. ECU developer

Restrictions on Diagnostic Access

The predefined access types and restrictions[54,55] for diagnostic access to ECUs in the operating system should not be disregarded and used exhaustively.

Important in all measures, however, is a reduction in possibilities. Limiting the possibilities refers to several aspects.

Mission 26 *Limitation of diagnostic access to **authorised users** and **safe conditions**.*

Restriction in Diagnostic Functions

It cannot emphasized enough: the best protection against attacks via diagnostics is to reduce the diagnostic options to the bare essentials[56].

Only functions that are not implemented cannot be misused. The decision on which features are mandatory must be coordinated with the users[57] of the diagnostics in critical dialogue.

Internal Diagnostics

ECUs with the ability of performing diagnostics[58,59] have a certain charm, they offer very interesting options for OEM[60], legislators[61] or insurances[62].

But a detailed analysis of these options discloses attack paths which have to be closed.

[54] see subsection 9.4.2 - Building barriers on page 106
[55] see Figure 65 - samples of diagnostic barriers on page 106
[56] see subsection 9.4.3 - Reduction of Diagnostic Options on page 107
[57] see subsection 9.1.1 - Users of Diagnostics on page 85
[58] telematics, emergency calls, infotainment etc.
[59] see subsection 10.3.5 - Teleservices on page 124
[60] failure analysis, update over the air
[61] monitoring of regulations (e.g. exhaust gas)
[62] monitoring driving behavior

Mission 27 *Definition and* ***limitation of abilities*** *of vehicle internal diagnostic tester.*

Routing Diagnostic Communication

In addition to the capabilities of the internal diagnostic tester as well as the routing of the diagnostic communication plays an important role. Attack using the diagnostic paths can also be limited by minimisation of signal routing[63].

Preventing diagnostic communication with a certain content to critical ECUs could be a very substantial protection.

Mission 28 *Implementation of firewalls for diagnostic communication in gateway ECUs.*

[63] see subsection 10.4.3 - Vehicle to Internet on page 126

12 Commercial Consequences

Safety and cost are not necessarily conflicting goals. Certainly, complex features are per se cost driver's which get an extra boost by safety and security demands. Decision-makers should be aware of that.

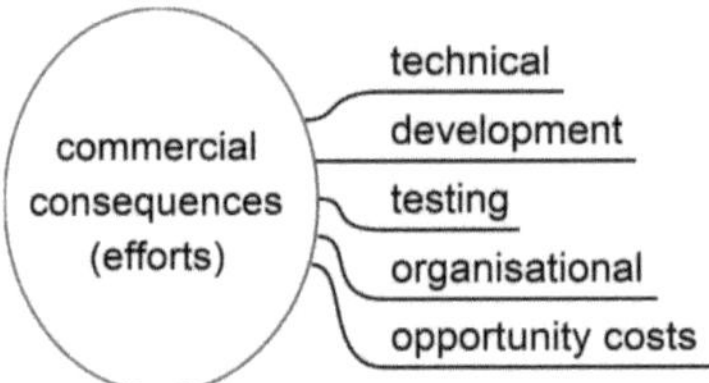

Fig. 77.: commercial consequences of increasing demand for security

Savings with a simultaneous increase of features are downright negligent. Support in decision-making should be a first line-up of expenses. This first draft has no claim to completeness, but it would be worthy to be analysed in greater depth.

12.1 Time Effort

It may sound quite strange, but technical innovations need a difficult reducible amount of time from idea to implementation, a fact that is just often overlooked by management[1].

Development work is brainwork. This means, conversely, you can indeed work faster, but do not think faster.

Mission 29 *Personnel capacities behave not proportional to the processing speed.*

[1] desire for first-to-market, idea to save development costs by speed

Increasingly sophisticated assistance systems require a more complex infrastructure. This increased complexity has to be somehow coped and ultimately tested.

Therefore well structured[2] and easy understandable[3] development processes and schedules[4] with sufficient reserves are imperative.

Already by the increased complexity of the functions and dependencies across can be expected with increasing safety[5] and security[6] demands, which requires compliance with certain standards[7] and resulting additional tasks[8].

Mission 30 *A function verification (A-sample) does* ***NOT*** *constitute completed development.*

Important - the same rules that apply to the OEM must also be claimed by the system supplier, but also conceded.

Mission 31 *It applies the Pareto principle. 80% of the work is done in 20% of the time. The remaining 20% really* ***need*** *the remaining 80% of the time. No false anticipation for rapid progress of the project just at the beginning!*

12.2 Technical Effort

Technical measures for hacking prevention does not come for free. The aspects are manifold. Features themselves require further hardware and/or software with protective measures for safety and security on top.

[2]following the organisational structure of the OEM and the supplier

[3]easy language, clear instructions

[4]product development plan

[5]functional safety

[6]network and system security

[7]e.g. ISO 26262

[8]e.g. additional analysis, further reviews, assessments, tests

There are no real surprises by checking the aspects. Maybe the extent is different[9] than expected.

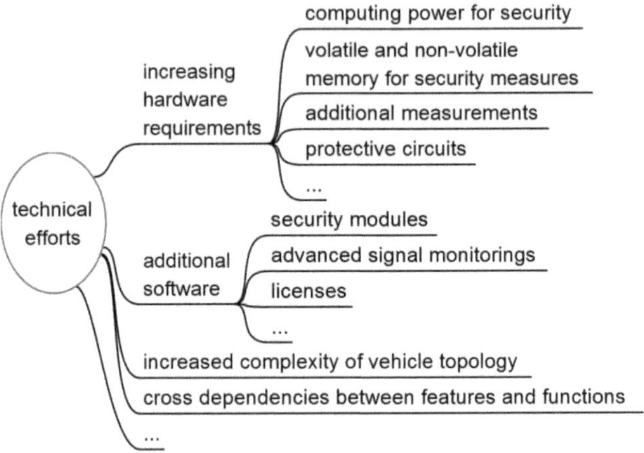

Fig. 78.: technical efforts

The estimation of the technical effort is not necessarily carved in stone. Especially engineers tend to consider technical requirements and resulting expenses as indispensable. So over-engineering is quite common.

Mission 32 *Reduction to the technical essentials, delimitation between usefulness and gimmick.*

Intensive interactions between decision-makers, engineering[10], safety and security department, supplier, purchase department and legislator are absolutely necessary. Finally, some decisions may be rather political than technical.

Mission 33 *Introduction of an iterational process for definition of technical comprehension regarding security with decision-makers, engineering and purchase department.*

[9] mostly higher in opting for a feature

[10] including R&D, production line, customer service, risk management

12.3 Development Effort

The causal relationship between feature and development costs should be logically comprehensible. However, some pitfalls may be still found.

The additional efforts for dealing with the increased complexity, coordination, communication and significantly much more detailed risk analysis are often underestimated. The relationship between feature and effort is definitely not linear.

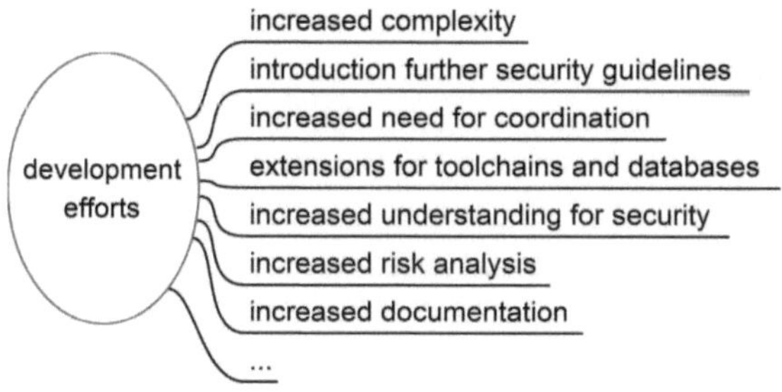

Fig. 79.: development efforts

Mission 34 *Sufficient consideration of incremental costs by increased* ***complexity****.*

Additionally, requirements for safety and security generate costs which cannot be allocated to individual projects[11] such as development of supplementary cross-sectional specifications or guidelines.

Mission 35 *Consideration of necessary cross-project activities.*

12.4 Testing Effort

Significantly underestimated is often the effort for testing activities in general. Safety and security topics are enlarging this problem additionally.

[11] vehicle projects, platform projects

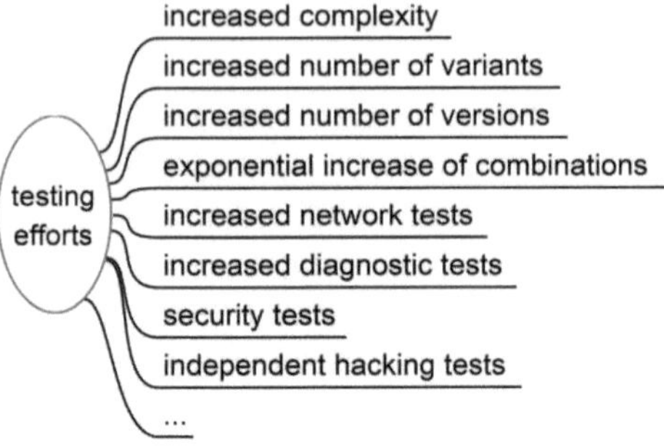

Fig. 80.: testing efforts

A particular cost driver for the testing is the increasing number of variants and the resultant combinatorics.

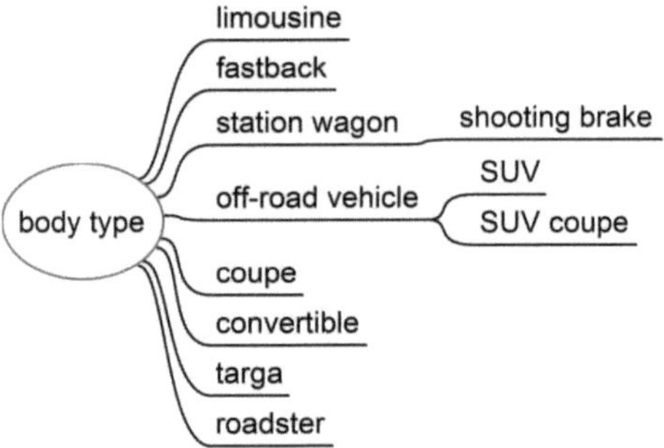

Fig. 81.: variants of car body types

In principle, each combination of separately offered features should be tested at least for each platform, vehicle model, sometimes even each derivate or body type[12] depending on their influence on the function.

Mission 36 *Including testing aspects in variant planning.*

[12] Figure 81

To the body types the various engine[13], transmission[14] and drive variants[15] and optional extras can now be added.

The test methods and contents are extremely diverse even without consideration of hacking aspects. Many tests are nevertheless quite systematic.

Tests for vulnerabilities that can be misused by hackers are only partly systematically. Most of the tests require a high degree of creativity instead. Furthermore, they require considerably more knowledge than automotive know-how.

It is assumed that some testing laboratories[16] will specialize on these tasks and will require even higher hourly rates.

Mission 37 *Considering necessary degree of specialisation and rising hourly rates.*

12.5 Organisational Effort

The high degree of specialisation regarding security sets demands on the organisation structure of all participating companies.

So more and more security experts should be trained and positioned in various departments, both speciality departments[17] as well as back-office functions[18].

Depending on the thematic orientation the experts should have specific tasks.

[13]petrol or diesel engine, hybrid drives, electrical engines

[14]manual gear, dual clutch transmission, continuously variable transmission (CVT), automatic gearbox with torque converter

[15]front wheel drive, rear wheel drive, 4-wheel drive(s)

[16]may also be prescribed by law

[17]infotainment electronics, chassis electronics etc.

[18]networking department, diagnostics department, functional safety department, testing department

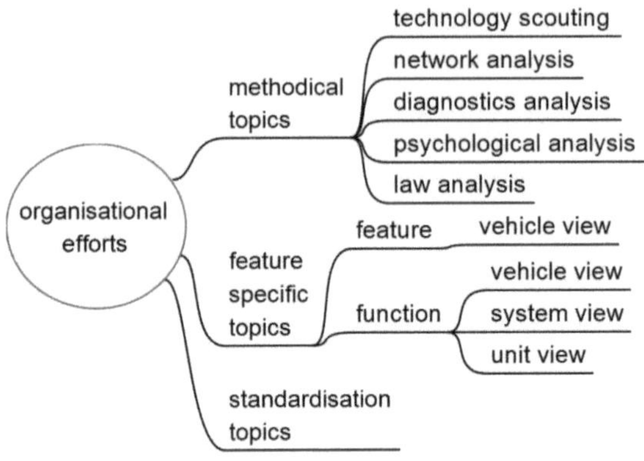

Fig. 82.: organisational efforts

Cross-sectional topics, even in unexpected combinations[19], should be either integrated into existing divisions at high specialisation or combined in a central, superordinate department for general, multidisciplinary topics.

Mission 38 *Positioning of experts in corresponding departments.*

In addition to professional counseling for decision-makers, marketing, designers, purchase department etc. the security department(s) should be also liable for R&D releases.

Mission 39 *Extending the powers of the security department to release relevance.*

Among the organisational efforts also includes a reconsideration of data security of IT, especially the areas that are required for the production[20] and customer service[21].

---

19 e.g. security and psychology

20 database of produced vehicles

21 online syncronisation of hardware and software versions of vehicles in the field

The number of information collected and stored in databases of production line and customer service increased in the recent years substantially.

Further on in the customer service process there are data of the vehicles life cycle[22] collected. Partly it contains quite personal data[23,24] which should be protected by data privacy [22].

Regardless of the sense and nonsense of the collection of such data, security and access restrictions defined in the IT must be implemented.

Attacks on these data bases and diagnostic interfaces by organised crime or secret services would raise hacking[25] to a completely new quality.

Mission 40 *Analysis of the collected data protection of vehicles and driver's behaviour in the IT.*

[22] e.g. mile age, maximum high speed

[23] e.g. geographical positions, last 10 parking positions

[24] see subsection 3.1.3 - Privacy and Security on page 32

[25] data theft, enabling of unauthorised changes

13 Opportunity Costs

The commercial implications of security are supposed to lead directly to a consideration of opportunity costs. Strangely enough, this consideration is hired relatively hardly.

Fig. 83.: opportunity costs

The ideas of this term[1] [49] are quite old[2], but more current. The phrase "Time is Money" by Benjamin Franklin [45] is cited very often and belongs into the midst of the context.

While the theory of opportunity costs only makes a distinction between explicit and implicit costs, a further prefixed distinction into monetary and non-monetary terms is quite reasonable.

13.1 Monetary Aspects

The theory of opportunity costs, rooted in microeconomics, is dealing mainly with monetary, financially clearly assessable values.

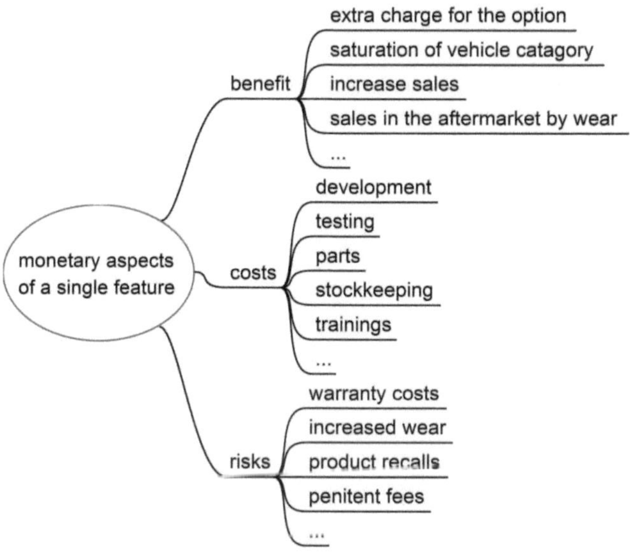

Fig. 84.: monetary aspects of a single feature

Looking at the benefit and cost situation in addition to the risks of a single feature results in interesting questions.

[1] Friedrich von Wiesner (1851-1926)

[2] based on theories of Benjamin Franklin (1706-1790) and Frédéric Bastiat (1801-1850)

- Are really **all costs** considered sufficiently?
- Are really **all risks** considered sufficiently?
- Are **costs and risks** really compensated by the benefit?

Finding the answers requires not only a structured approach, but also a great honesty without consideration of professional sensibilities[3].

Mission 41 *Unbiased capturing and comparison of all costs, risks and benefits.*

13.2 Non-Monetary Aspects

More difficult to grasp but likewise important are the non-monetary aspects. These considerations are company policy.

Many of these considerations are more or less secret and are therefore hardly or not communicated. Finally, the visibility, the reputation and the credibility of the company could be damaged easily.

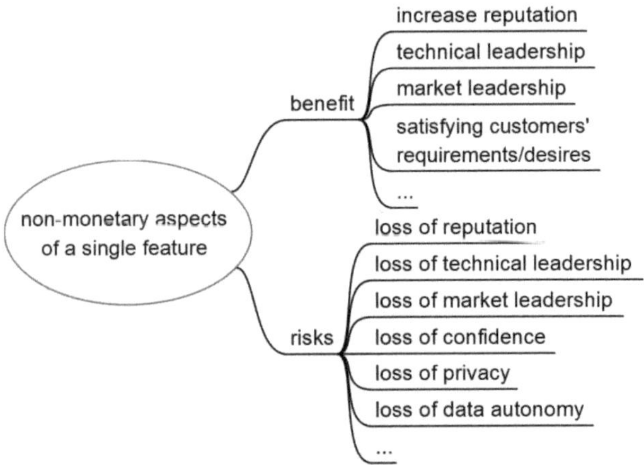

Fig. 85.: non-monetary aspects of a single feature

[3] visions of design and marketing, playfulness of engineers, fears and distrust of quality assurance etc.

Non-monetary benefits and risks should be named[4] and weighed.

A straightful handling with these issues facilitates decision making considerably and creates transparency for everyone involved.

Mission 42 *Identification, naming and weighing of non-monetary benefits and risks.*

[4] samples see Figure 85

Summary

The technology of the future has as wonderful as dangerous possibilities, but resignation is not necessary. Various measures to avoid risks of a destructive handling of the new technologies are already in use.

The challenge will now be to search vulnerabilities using systematic methods and destructive ideas and eliminate them sustainably.

To avoid hacking in the future will be impossible. So the ultimate goal should be to make attackers life as difficult as possible. At this goal is necessary to work and research together across disciplines.

List of Figures

List of Tables

Bibliography

[1] *Adaptive Cruise Control and Beep of Impending Doom.* http://www.caranddriver.com/features/five-annoying-safety-technologies-adaptive-cruise-control-and-beep-of-impending-doom-page-3

[2] *The AIRMATIC air suspension system. - Mercedes-Benz.* https://www.mercedes-benz.com/en/mercedes-benz/innovation/the-airmatic-air-suspension-system/

[3] *Android.* https://www.android.com/

[4] *Arduino - Home.* https://www.arduino.cc/

[5] *Audi magnetic ride – Audi Technology Portal.* http://www.audi-technology-portal.de/en/chassis/suspension-control-systems/audi-magnetic-ride_en

[6] *BeagleBoard.org - bone.* http://beagleboard.org/bone

[7] *BMW cars found vulnerable in Connected Drive hack | PCWorld.* http://www.pcworld.com/article/2878437/bmw-cars-found-vulnerable-in-connected-drive-hack.html

[8] *BMW i Vision Future Interaction auf der CES 2016: Offener i8 zeigt Cockpit der Zukunft (Bildergalerie, Bild 13) - Auto Motor und Sport.* http://www.auto-motor-und-sport.de/bilder/bmw-i-vision-future-interaction-auf-der-ces-2016-4637528.html?fotoshow_item=12

[9] *BMW i8 Mirrorless auf der CES 2016: Wir haben den spiegellosen i8 ausprobiert (Bildergalerie, Bild 12) - Auto Motor und Sport.* http://www.auto-motor-und-sport.de/bilder/bmw-i8-mirrorless-ohne-spiegel-ces-2016-10343810.html?fotoshow_item=11

[10] *BMW steps into augmented reality with AR driving glasses for Mini.* http://mashable.com/2015/04/19/bmw-mini-ar-driving-glasses/?utm_cid=mash-com-pin-link#M_2Br7ZbZPqX

[11] *BMW zeigt "ConnctedRide" auf der CES 2016: Laserlicht und Cyborg-Helm mit Head-Up-Display (Bildergalerie, Bild 5) - Auto Motor und Sport.* http://www.auto-motor-und-sport.de/bilder/bmw-ces-2016-connected-ride-helm-head-up-display-laserlicht-10343866.html?fotoshow_item=4

[12] *CAN-Bus Data Capture with Wireshark on Raspberry Pi | SK Pang Electronics Ltd.* http://skpang.co.uk/blog/archives/1141

[13] *Carberry.it.* http://www.carberry.it/

[14] *Carlsson Autotechnik - Mercedes Tuning auf Gut Wiesenhof - Aerodynamik, Tagfahrlicht , Leistungssteigerung, Räder, Felgen.* http://www.carlsson.de/carlsson/de/Produkte/C-Tronic/C-Tronic.php

[15] *CES 2016: Toyota verspricht unfallfreie Autos dank Künstlicher Intelligenz | heise online.* http://www.heise.de/newsticker/meldung/CES-2016-Toyota-verspricht-unfallfreie-Autos-dank-Kuenstlicher-Intelligenz-3062821.html

[16] *CES: Nvidia stellt Rechnersystem für selbstlenkende Autos vor | ZDNet.de.* http://www.zdnet.de/88255945/ces-nvidia-stellt-rechnersystem-fuer-selbstlenkende-autos-vor/?utm_source=rss&utm_medium=rss&utm_campaign=rss

[17] *DIY Remote control Casing - RobotShop Forum.* http://www.robotshop.com/forum/diy-remote-control-casing-t7115

[18] *Electric Park Brake | ZF TRW.* http://www.trw.com/braking_systems/electric_park_brake

[19] *Electric power steering Servolectric Bosch Automotive Steering.* http://www.bosch-automotive-steering.com/en/products/car-steering-systems/electric-power-steering-servolectric.html

[20] *Elektrik/Elektronik – Audi Technology Portal.* http://www.audi-technology-portal.de/de/elektrik-elektronik

[21] *The Eponymous Pickle: Stylish BMW AR Goggles for in Car.* http://eponymouspickle.blogspot.de/2015/05/mini-bmw-ar-goggles-for-in-car.html

[22] *FIA deckt auf: BMW speichert Daten - Auto Motor und Sport.* http://www.auto-motor-und-sport.de/news/fia-deckt-auf-bmw-speichert-daten-8428028.html

[23] *GENIVI Alliance.* http://www.genivi.org/

[24] *HOME - Ingenic.* http://iwop.ingenic.com/en/

[25] *Home : AUTOSAR.* http://www.autosar.org/

[26] *How activity trackers remove our rights to our most intimate data | Technology | The Guardian.* http://www.theguardian.com/technology/2014/jun/03/how-activity-trackers-remove-rights-personal-data

[27] *Jaguar Concept Windshield Shows Off Augmented Reality in the Car.* http://mashable.com/2014/07/14/jaguar-augmented-reality/#wxNoSt9yp5q_

[28] *Jeep Hacking 101 - IEEE Spectrum.* http://spectrum.ieee.org/cars-that-think/transportation/systems/jeep-hacking-101

[29] *Mercedes-Benz Deutschland.* http://t.mercedes-benz.de/content/germany/mpc/mpc_germany_website/de/home_mpc/passengercars/mobile/mbot/new_cars/model_overview/c-class/w205/pictures/pictures_videos.html

[30] *ND-BC4 - Universal Rear-View Camera | Pioneer Electronics USA.* http://www.pioneerelectronics.com/PUSA/Car/Accessories/Rear-View-Cameras/ND-BC4

[31] *New Industrial Protocols modules: RS-232, RS-485, CAN Bus, Modbus, 4-20 mA for Arduino, Raspberry Pi and Intel Galileo / Cooking Hacks Blog.* https://www.cooking-hacks.com/blog/new-industrial-protocols-modules-rs232-rs485-can-bus-modbus-4-20ma-for-arduino-raspberry-pi-intel-galileo/

[32] *The number of Keyless cars thefts is risingSecurity Affairs.* http://securityaffairs.co/wordpress/29672/cyber-crime/number-keyless-cars-thefts-rising.html

[33] *ODROID | Hardkernel.* http://www.hardkernel.com/main/main.php

[34] *OXY®your daily Companion |.* http://www.oxytechs.com/en-US#

[35] *Raspberry Pi - Teach, Learn, and Make with Raspberry Pi* https://www.raspberrypi.org/

[36] *Smartwatches: Stiftung Warentest vergleicht cookoo, i'm Watch, Pebble, Galaxy Gear und SmartWatch 2 - Notebookcheck.com News.* http://www.notebookcheck.com/Smartwatches-Stiftung-Warentest-vergleicht-cookoo-i-m-Watch-Pebble-Galaxy-Gear-und-SmartWatch-2.107883.0.html

[37] *Tizen | An open source, standards-based software platform for multiple device categories.* https://www.tizen.org/

[38] *Ultimate Single Board Mini PC for Android and Linux - UDOO*. http://www.udoo.org/

[39] *What happens when Tesla's AutoPilot goes wrong: owners post swerving videos / Technology / The Guardian*. http://www.theguardian.com/technology/2015/oct/21/tesla-autopilot-goes-wrong-videos

[40] *What's behind Toyota's big bet on artificial intelligence?* http://www.autonews.com/article/20151109/OEM06/311099937/whats-behind-toyotas-big-bet-on-artificial-intelligence?

[41] *Wi-Fi Alliance*. http://www.wi-fi.org/

[42] *Wi-Fi Aware / Wi-Fi Alliance*. http://www.wi-fi.org/discover-wi-fi/wi-fi-aware

[43] *Wikipedia*. https://www.wikipedia.org/

[44] *Moore's law*. https://en.wikipedia.org/w/index.php?title=Moore%27s_law&oldid=686725925. Version: Oktober 2015. – Page Version ID: 686725925

[45] BENJAMIN FRANKLIN: *Advice to a Young Tradesman*

[46] CEZARY KASPRZAK: *The Influence of Infrasounds on the Electrocardiograph Patterns in Humans*. http://przyrbwn.icm.edu.pl/APP/PDF/118/a118z1p20.pdf

[47] DÖLLE, Mirko: Autoschnüffler. In: *c't* 25 (2015)

[48] FASTL, Hugo ; ZWICKER, Eberhard: *Psychoacoustics: facts and models ; with ... 53 psychoacoustics demonstrations on CD-ROM*. 3. ed. Springer (Springer series in information sciences 22). – ISBN 978–3–540–23159–2

[49] FRIEDRICH VON WIESNER: *Theorie der gesellschaftlichen Wirtschaft*

[50] GLEICH, Clemens: Kontrollverlust am Steuer. In: *c't* 24 (2015)

[51] GUDERA, Andy: *Diagnose-Handbuch - Allgemeine Grundlagen der Fahrzeugdiagnose (internal document)*. Berner & Mattner Systemtechnik GmbH, 2015

[52] LEHNERT, Peter ; SCHMIDT-CLAUSEN, H.-J: *The effect of the vehicle dynamics on the light distribution of headlamps*

[53] RÜSBERG, Kai: Keyless gone. In: *c't* 26 (2015)

[54] SUROWIECKI, James: *The wisdom of crowds.* 1. ed. Anchor Books. – ISBN 978–0–385–72170–7

Index